THE SCIENCE AND ETHICS
OF ENGINEERING
THE HUMAN GERM LINE

Mendel's Maze

T0324852

THE SCIENCE AND ETHICS OF ENGINEERING THE HUMAN GERM LINE

Mendel's Maze

JON W. GORDON, M.D., Ph.D.

Department of Geriatrics and Adult Development
Mount Sinai School of Medicine
New York, New York

WILEY-LISS

A JOHN WILEY & SONS, INC., PUBLICATION

Library of Congress Cataloging-in-Publication Data:

Gordon, Jon W., 1949–
   The science of genetic and reproductive technol0gies / Jon W. Gordon.
      p. ; cm.
Includes bibliographical references and index.
      ISBN 0-471-20647-4 (cloth : alk. paper)
   1. Genetic engineering—Popular works.   2. Human reproductive
technology—Popular works.
   [DNLM: 1. Genetic Techniques.   2. Reproductive Techniques. WQ 208
G663s 2004]   I. Title.
   QH442.G67 2004
   660.6'5—dc21   2003009970

10 9 8 7 6 5 4 3 2 1

This book is dedicated to two mentors who have contributed immeasurably to my intellectual and scientific development, and who have shown, by example, how to maintain the highest standards of scientific research and ethical conduct. First, to Clement L. Markert, my Ph.D. advisor, whose greatest gift to me was an unfettered and unrestricted access to his own titanic intellect; and, second, to Frank H. Ruddle, Professor of Biology, Yale University, who provided the resources and intellectual environment needed for my postdoctoral research, which led to the first successful introduction of foreign genes into the mouse germ line.

CONTENTS

PREFACE

As we all know, we are in the midst of a revolution in biomedical science. As I discuss later in this book, most of us labor under the illusion that this revolution has arrived precipitously and is the result of a few scientific giant steps. The most obvious of these startling steps forward at the moment is the cloning of the sheep Dolly in 1997. This breakthrough astounded most scientists, who were familiar with the enormous difficulties encountered in decades-long attempts to clone frogs, and the lay public, which was suddenly confronted with a procedure that could readily be applied to humans for production of innumerable genetic replicates of any individual. But, in contrast to the popular perception that mammalian cloning has suddenly thrust us to the threshold of a "brave new world" of reproductive medicine and genetics, it has, in fact, been a steady advance and convergence of many previously unconnected scientific disciplines that has brought us to the point of considering the pros and cons of human genetic engineering.

What are these unrelated disciplines? Very important among them is recombinant DNA technology. The ability to isolate, purify, and amplify individual genes, express those genes in bacteria and cultured cells, splice those genes so as to create "expression cassettes" with novel properties, insert new genetic material into cultured human cells and embryos of a variety of mammalian species, and efficiently determine the entire sequence of the genetic material in humans has been a critical component of this provocative progress toward human germ line genetic manipulation. It is worth noting that the implications of recombinant DNA technology for advancing human biology and medicine and for creating unsettling new opportunities for both understanding and modifying the lexicon of human genes was not lost on its original developers. In 1975, shortly after viral genes were first cloned in bacteria, a conference was held in Asilomar that both called attention to these new potentials and discussed the possible risks involved. However, other advances in biomedical science have mandated that these issues be considered in a new context.

Very important among these other advances has been progress in reproductive medicine. The birth of Louise Brown, whose development began in a culture dish during the process of in vitro fertilization (IVF) in 1978, established a strategy for obtaining large numbers of human embryos, manipulating them in the laboratory, and transferring them back to women for subsequent development to term. IVF technology has been further potentiated by embryo freezing as well as the development of procedures for achieving fertilization by manual insertion of a single spermatozoon into the egg. Reproductive technologies that are closely related to those used in IVF and that have thus far been used only in animals have also progressed to the point where they could be applied to human genetic engineering. These include embryo splitting, embryo aggregation, and the production of clones by transfer of nuclei from embryos, fetuses, or adults into eggs.

Finally, we have acquired a far more comprehensive understanding of complex developmental processes. Clusters of genes that control development of intricate structures—the so-called homeobox gene clusters—were first identified in the fruit fly *Drosophila,* and the hierarchical mechanisms of control of genes within these clusters have largely been elucidated. Corresponding and related gene clusters have also been identified in mammals, as have genes that regulate the process by which cells become committed to certain pathways of specialization.

So, these advances in recombinant DNA technology, reproductive medicine, reproductive biology developmental biology, and to some extent, genetics, have brought us to the point of considering the feasibility and advisability of modifying the human germ line. Spectacular developments like the birth of Dolly serve as visible "road signs" on the route to human germ line manipulation rather than the destination itself.

It is hardly surprising that the public has not maintained a high awareness or ongoing interest in the slow and steady developments of these many areas of science and medicine. However, the relatively recent coalescence of these disciplines has had another significant consequence: their interdependence has likewise not been apparent to many specialists in the respective scientific and medical fields. As a consequence, gene transfer experts hold symposia in which they describe strategies for human genetic engineering without understanding that these strategies must be pursued within the context of the proper practice of clinical medicine. Some members of the lay public, and even some obstetricians, have expressed the intention of performing human cloning without an awareness of findings from developmental biologists and molecular biologists that clearly indicate that human cloning would likely have disastrous consequences for the fetus as well as the woman who carried it. Molecular biologists, reproductive biologists physicians and, to some extent, bioethicists, talk about human genetic engineering without an awareness of findings from developmental biologists, which indicate that rigorous control of human development would be exceedingly difficult to achieve. The lack of a holistic and comprehensive understanding of human genetic manipulation on the part of the public as well as groups of specialists has led to a disorganized and often misleading discussion of the issue. One major goal of this book is to demonstrate the importance of recognizing the interrelationship of these scientific and medical fields when evaluating the appropriateness of genetic modification.

How are members of the lay community who, in a democratic society, will ultimately be responsible for determining how human genetic engineering is to be regulated, to cope with this barrage of scientific and medical breakthroughs, and with the profound ethical questions posed by the specter of a human genome redesigned? First, they must be provided with basic scientific and medical knowledge needed to understand genetic manipulation. This core of knowledge need not include details of molecular structure, X-ray crystallography of nucleic acid or protein structure, fine details of human anatomy, or an extensive discussion of the philosophical underpinnings of modern day bioethics. Rather, it must contain a practical distillation of these details that provides a contextually relevant and functional understanding of the biological and physical mechanisms that govern clinical applications of genetic engineering technology, as well as an acquaintance with principles of medical ethics that certainly should govern those applications. My firm belief is that this information can be readily absorbed by persons with no direct familiarity with these issues, and production of this book is a reflection of that conviction.

In producing this book, I have other objectives additional to that of providing a clearer and more encompassing appreciation of the potentials and limitations of human germ line gene modification. One of these is to provide a plan, or recipe, if you will, for analyzing and judging the suitability, for human use, of new technologies as they appear on the scene. It is my hope the use of this analytical protocol will ease the shock waves as we are confronted by each new ethically challenging forward step toward genetic engineering. The other purpose of the plan is to encourage retention of a disciplined approach to analysis of new breakthroughs. We must not lose sight of the principles upon which we wish to guide our acceptance of new technologies and, at the same time, we need resist the urge to factor into logical analysis fundamental philosophical issues that are themselves impossible to resolve through logic or fact gathering. This is not to say that our fundamental concepts of right and wrong are not important for the process of judging new capabilities in the area of genetic engineering, but rather, to indicate that these concepts and principles, as we see them, must assume their proper role in the reasoning process.

My third goal is to provide an appreciation of the importance our social history has in influencing our analytical methods. Most of us are fond of believing that our reasoning is both enlightened and free from biases. These assumptions are, of course, completely false. No person lives in a social and historical vacuum, and thus, all of our thinking is influenced by biases, be they conscious or, more likely, below the level of our awareness. Where the subject of human genetic manipulation is concerned, our biases concerning the control women ought to have of the reproductive process are very important. No genetic manipulation procedure can be undertaken without the participation of women as carriers of embryos and fetuses. This means, of course, that women must not only lend their wombs to the effort, but they must endure the significant physiological and psychosocial impact of a potentially risky pregnancy. It is no small matter to make such an investment, and it is my hope that as we move toward practical implementation of genetic engineering, we will not lose sight of the fact that the physical and emotional integrity of the women who must inevitably take all of the risks should be protected and preserved to the greatest extent possible.

Much of the discussion of human genetic manipulation has had an ominous tone. Are we going to redefine what in means to be human? Are we going to alter the process of evolution? Are we going to create monsters? Are we playing God? As I hope this book will show, these more profound and disturbing prospects are less likely to become realities if we reaffirm the principles upon which we assert that our free, democratic system is built. We must maintain our commitment to individual freedom, self-determination, sovereign control over one's physical person, and fundamental equality. If we keep these overarching principles in sight and act to reinforce them, human germ line genetic manipulation, to the extent it is effective, will be all the more likely to be a beneficial addition to our armamentarium of methods for treating and preventing painful and debilitating disease.

ACKNOWLEDGMENTS

I would like to thank the following individuals for helpful discussions and/or assistance with background research: Daniel Gordon, Jacob Gordon, Nancy M. P. King, and Marta Rico. Special thanks to Sue Ann Fung-Ho, Media Resource Center, The Rockefeller University, New York, NY for assistance with the artwork.

PART I

1

SETTING THE TABLE

Life is short, and the Art long; the occasion fleeting; experience
fallacious, and judgment difficult
—Hippocrates, 470–410 BC

The best and socially most acceptable way to enjoy an order of french fries is to snatch them up with a clean pair of hands and stuff them into your mouth. But what about a dinner of vichyssoise, escargots, roast duckling a l'orange, pureed parsnips, new potatoes sautéed in clarified butter, a glass of Chardonnay, and crêpes Suzette? To fully enjoy a repast like this one, the table must be properly set.

An examination of the issues surrounding human germ line genetic modification is an intellectual gourmet meal. The diverse scientific, technological, medical, and philosophical disciplines that bear on this emerging prospect are intellectually as exciting and rewarding to explore as are the gustatory stimuli of a duckling l'orange with all the fixings. However, a coherent view of this new frontier in biomedical science cannot be attained by examining any one of the related issues in isolation, just as those escargots would be incomplete if served à la carte. Moreover, if it were necessary to shovel the parsnips into one's mouth with cupped fingers, some of the more fulfilling experiences of that gourmet dinner would be ruined. What is needed to properly assess and judge the desirability of human germ line modification is a harmonious and holistic consideration of all aspects of the problem, using the appropriate tools for analysis. Our intellectual "fork and knife" must be a familiarity with the relevant academic disciplines.

The Science and Ethics of Engineering the Human Germ Line: Mendel's Maze, by Jon W. Gordon
ISBN 0-471-20647-4 Copyright © 2003 John Wiley & Sons, Inc.

Discussions of the prospect of human germ line genetic manipulation, reinvigorated almost daily by reports of new scientific breakthroughs, have heretofore largely suffered from a failure to consider all of the relevant issues in an encompassing approach and/or from a deficiency in the basic scientific or medical knowledge (the deficiency often being greatest in those voicing their points of view most vociferously) required to arrive at a cogent analysis of the issues. Of course, it's both easy and fun to imagine futuristic scenarios in which institutes of physics are staffed entirely by individuals with capacities of Einstein, or where a basketball team has five clones of Michael Jordan. I would be the last person to deter the reader from such entertaining diversions. However, when speculation is confused with calculation and inappropriately used to make policy or develop legislation, the risk becomes significant that people with real medical needs will be deprived of new advances in medical science or that their freedom to make medical choices and maintain the confidentiality of those choices will be unfairly curtailed. Another and perhaps most ominous consequence of such undisciplined thinking is that unrealistic doomsday scenarios, mistaken for measured prognostications of the future, can create pressure to restrict basic research and discourage free inquiry. It therefore behooves us to recognize and distinguish fun and frolic from formal analysis.

One of the conundrums of our democratic society is that while the vitality of our system depends on the participation and input of the people at large, many of the complex issues faced in this modern world seem beyond the expertise of the average person. As a result, we tend to leave major decisions up to "the experts." The consequence of this reality is that vocal minorities often have disproportionate influence on policy making and that the vested interests of "the experts" are often served at the expense of the public good. A meaningful understanding of human germ line genetic manipulation certainly appears, at first glance, to require a knowledge of science, medicine, and bioethics that is unattainable for the average person, given the limited time and energy available to those dealing with the immediate demands of daily life. Yet in almost no other major social arena is it more important that the general public, whose very concept of what it is to be human may be impacted profoundly by the use of this technology, and whose society may be fundamentally and irreversibly changed by it, contribute to a discussion of the desirability of deliberate manipulation of the human genome.

It is my contention that the knowledge required to render a well-reasoned judgment of the appropriateness for human use of each new advance in reproductive genetics is readily attainable for any literate person. With a few analytical principles in hand, and with a knowledge of some basic scientific facts, it is quite possible for an individual without an advanced knowledge of biology, medical technology, or bioethics to fully comprehend, appreciate, and judge the merits of extant and future methodologies that have the potential for deliberate manipulation of the human genome. On the scientific side, the information required is not only straightforward but wonderfully fascinating and exciting. On the medical side, what must be attained is a perspective on provocative new technologies that does not lose sight of the inviolable principles that should govern all patient-doctor relationships. In the area of bioethics it is necessary to recognize the hierarchical nature of the subject and to determine whether each question within its purview is amenable to logical analysis or falls instead within the area of philosophical or religious dogma. Finally, it is important to appreciate that we do not live and think in an historical vacuum. Much as we may wish to deny it, we all have prejudices and biases—legacies of our social and personal histories. These biases are far less likely to hamper our objectivity if they are openly acknowledged.

I am certain that the information essential for attaining the level of understanding of human genetic engineering needed to cogently evaluate each new scientific and medical breakthrough, as it brings such procedures ever closer to technical feasibility, can be attained, and it is the purpose of this book to provide that understanding. Before we embark on a detailed discussion of human germ line manipulation, it is useful to outline the areas of science, medicine, and ethics with which we must be familiar to be prepared to evaluate this emerging technology.

Science

Several fundamental principles of biology must be understood before "genetic engineering," with all of its ramifications, can be understood. We must appreciate the relationship between the structure and function of the chemical compounds contained within a living organism and the life process. That is, we must recognize that a living being is a biological machine, built meticulously from the fertilized egg, conforming to the laws of chemistry and physics, and maintaining itself in accordance with instructions contained within an information data bank. Although the biological machine that we perceive as a living organism is far too complex to understand completely at the present time, the structure-function relationship in biological systems can be readily grasped and applied to an analysis of genetic engineering. It should be emphasized at the outset that the notion of living creatures as biological machines in no way negates the Judeo-Christian concept of the "soul." The soul is a metaphysical concept and is presumed to be an entity entirely distinct from the body; thus an examination of the mechanics of body construction and function are separable from our understanding of each "soul" as unique and irreplaceable. Indeed, most of us already agree that genetic clones such as identical twins or artificially constructed clones that might be produced in the future are each unique individuals. Therefore, a discussion of the biological machine is neither an endorsement nor a refutation of the metaphysical concept.

Machines are of course made of parts, and we will examine the structure and function of the parts of the biological machine in the degree of detail that is necessary to appreciate how genetic modification is likely to alter the construction process and, ultimately, the final product. The basic unit of information is the gene, and we must understand what genes can and cannot do if we are to appreciate the potentials and limitations of genetic modification. Of course, a complete understanding of the relationship of gene function and human development is not yet in hand—it remains one of the compelling problems of science. But we do know enough at this time to attain a pretty good understanding of what we can and cannot control about the relationship between "genotype" (our endowment of genes) and "phenotype" (the final result of our development from a fertilized egg to an adult). As we evaluate efforts to manipulate the human genome, it is just as important (perhaps more important!) to recognize what we do not know, and what we may never know, as it is to understand what is known.

Understanding the relationship between gene function and human development requires an appreciation of the level within the structural hierarchy at which genes have their impact. To attain this appreciation, the relationship between gene structure and function must be understood. Genes are chemical compounds, of course, and they provide instructions for assembly of other chemical compounds that actually carry out the activities of cells, the basic functional subunits of the intact organism. The mechanics of this information storage and decoding must be clear to us if we are to predict whether modification of the genome will change the characteristics of the conceptus in the manner we choose. Exactly how genes regulate construction and operation of the biological machine is important if one is to evaluate the practicability of germ line gene modification. Therefore, we will examine the kinds of functions encoded by genes.

One of the great concerns of those alarmed by advances in genetic manipulation is that such interventions can change not only the direct recipient of the manipulation but also his or her offspring for generations to come. The fact that germ line genetic modifications are heritable means we must understand the mechanisms of inheritance if we are to assess the risks associated with changes that can be passed on to children. Only through this understanding can we distinguish the modes of inheritance that should be of concern from those that would be of minimal significance. Thus dominant, recessive, and sex-linked inheritance must be reviewed within the context of our understanding of how genes work at the molecular level.

A fact not adequately emphasized in most discussions of genetic manipulation is that all of such interventions require control of the human reproductive process. If sperm, eggs, or embryos are to be manipulated, fertilization and the early subsequent phases of development must take place outside the human body. Accordingly, genetic engineering depends for its success on efficient methods for obtaining sperm and eggs, producing healthy embryos in the tissue culture environment, and obtaining pregnancies once these embryos are returned to the female. Therefore, if we are to evaluate the feasibility and ethical acceptability of extant or proposed methods of genetic manipulation, we must have a good grasp of the limitations imposed by in vitro reproductive technologies, and we must also take into account the risks individuals might be required to take in order to produce and nurture embryos subjected to genetic manipulation. An introduction to reproductive biology and assisted reproduction technologies (so called "ART" technologies) is therefore needed.

Medicine

Often lost in discussions of cloning or related genetic manipulations is the realization that, when extended to humans, these procedures must be regarded as invasive medical interventions. When an individual is approached for medical therapy, or when an individual requests an elective procedure, many issues arise that are nonexistent when the same techniques are used in animals. Issues of risk-benefit ratio, cost, and treatment alternatives must be considered before human beings are manipulated to their potential peril. Moreover, recipients of potentially dangerous interventions must be informed of risks and must consent to those procedures in a setting that would satisfy a neutral observer that the treatment candidate is adequately informed. We cannot forget that those we approach for genetic manipulation must retain control of their fates as much as possible, for it is they who may suffer terribly if things go wrong. Thus we will discuss implementation of these technologies in the context of an effort to develop an appropriate informed consent process.

It is also necessary to keep in mind that even in today's relatively enlightened environment, the decision to perform a medical procedure is affected by social biases that have deep historical roots. If and when the human germ line is manipulated, most of the risks will be taken by women, and historically women have not enjoyed the same social status as men. We must guard against development of attitudes toward potentially dangerous procedures that are more relaxed because the procedures involve women. In addition, we must be cognizant of the potential social impact that limiting genetic engineering to the privileged few might have.

Ethics

Scientific information is simply factual and is consequently relatively easily transmitted and grasped. In the area of ethics and philosophy, however, absolute truths are less easily found. How can we possibly sort through the myriad ethical issues that arise when considering invasive medical procedures that can affect the health of generations yet unborn and confer fundamental advantages upon those who receive them? What principles can we apply to analysis of such questions that will assist us in assessing the desirability of human genetic manipulation?

As complex as these issues are, they can be evaluated in an organized way that is consistent with basic philosophical tenets of our society. To accomplish such an evaluation we must accept that reasonable people with the best intentions can disagree on the most profound questions in medical ethics. In The Netherlands, a country regarded as generally similar to our own in its moral outlook, assisted suicide is far more accepted than in the United States. Who's right? Of course, there is no "right and wrong" where such questions are concerned. What we must strive to do when we evaluate the prospect of human genetic engineering is to identify the ethical precepts on which there is unanimity and apply those precepts. As we sort through the various ethical issues raised by manipulation of the human germ line, we can apply those principles that we regard as inviolable, and in many cases arrive at clear decisions regarding the acceptability of a gene manipulation procedure without having to confront more fundamental philosophical questions for which no clear answer exists.

With this background we can now proceed to evaluate the science, medicine, and ethics of human gene manipulation. With the relevant scientific knowledge in hand we will see what is possible, what does not appear to be possible, and what new developments are likely to arise in the future. We will examine the related medical questions, questions that straddle the areas of science and ethics, and we will approach the ethical implications of this technology from a perspective that we hope will allow us to feel comfortable with our assessment of the various approaches to genetic manipulation that are likely to have the potential for effective use in humans. But first, the science.

2

BUILDING A LIVING ORGANISM FROM INANIMATE PARTS

What was life? No one knew . . . if there was anything that might
be said about it, it was this: it must be so highly developed,
structurally, that nothing even distantly related to it was present in
the inorganic world . . . it was not matter and it was not spirit, but
something between the two . . . like the rainbow on the waterfall. . .
. No doubt at all but just as the animal kingdom was composed of
various species of animals, as the human-animal organism
composed of a whole animal kingdom of cell species, so the cell
organism was composed of a new and varied animal kingdom of
elementary units, far below microscopic size, which grew
spontaneously, increased spontaneously and, . . . acting on the
principle of a division of labor, served together the next higher
order of existence. . . . For it was not possible to brush aside like
that the idea of the . . . rise of life out of what was not life. That gap
between . . . living and dead matter . . . must somehow be closed
up or bridged over. . . . Division must yield "units" which . . .
mediated between life and the absence of life . . . molecular groups,
which represented the transition between vitalized organization and
mere chemistry. . . .

—Thomas Mann, *The Magic Mountain,* 1927

What distinguishes a living entity from the vast majority of material in the universe, which is nonliving? An in-depth discussion of the evolution of life and

The Science and Ethics of Engineering the Human Germ Line: Mendel's Maze, by Jon W. Gordon
ISBN 0-471-20647-4 Copyright © 2003 John Wiley & Sons, Inc.

the conditions that fostered it is beyond the scope of this book. However, a few basic points that reinforce the notion of the living organism as a biological machine that obeys the laws of physics and chemistry are worth making.

The prescient passage from Thomas Mann quoted at the head of this chapter focuses on a key feature of life—its complexity. The cosmos, or "nature," can be reasonably viewed as a vast emptiness containing just a little bit of matter and energy. Moreover, inexorable and constant forces tend to disperse matter evenly throughout the universe. Consider a cube of solid iron, one mile wide, one mile deep, and one mile high, floating in deep space. The atoms in this cube are not motionless, because although deep space is very cold, the temperature is not absolute zero, a theoretical lower limit at which all motion ceases. Now consider the atoms on the very surface of this cube. On one side sits the remainder of the cube, while on the other sits the void of space. Forces of attraction tend to hold these surface atoms within the cube, but as time passes it is inevitable that an atom will break free from the cube and float into the adjacent space. Although such events may occur rarely, the amount of time available for the complete dispersion of the cube is unlimited. Therefore, over countless millennia the cube will gradually become a diffuse cloud of iron atoms, and over further expanses of time too great for most of us to comprehend the atoms from this cube will be distributed evenly across the universe. This tendency toward the randomization of matter, also known as the second law of thermodynamics or the law of entropy, is constantly at work, eroding objects and dispersing them throughout the cosmos. Now perhaps you may think that other constants, like gravity, can resist the law of entropy and provide for stable, organized structures. Stars like the sun have such enormous gravitational fields that they can hold objects like Pluto, nearly 4 billion miles away, in orbit. And what about the gravitational forces of major galaxies, which can be composed of more than one hundred billion suns? Although it is true that gravitational forces can hold objects together, it must also be appreciated that they can pull things apart. The sun is "tugging" on the earth constantly, exerting force that would tend to pull the earth apart. Therefore, on balance, in a presumably closed system such as the universe, randomness will eventually prevail. An analogy I find useful for the perfectly equilibrated cosmos is that of a body of water with a smooth surface. Depressions in the "sea of entropy" may be considered much like small vortexes in a liquid surface. These vortexes are not objects in and of themselves; they are transient distortions in the surface—concentrations of nonrandomly distributed matter that are created with energy input and that dissipate with time. They may be regarded as pockets of "negative entropy."

The enormous degree of complexity of a living organism can be appreciated by considering it as a depression in the surface of the sea of entropy. If we look at the known universe carefully, we are stunned by its simplicity. Most of the universe is space, and the vast majority of the matter contained within that space is extraordinarily simple: 99.99 percent of the known matter in the universe consists of hydrogen, the smallest atom, and helium, the second smallest atom. Stars are made up almost entirely of these two simple elements, but as huge conglomerations of these elements they are sizable pockets of negative entropy. Galaxies, which contain billions of stars, are magnificent depressions in the smooth surface of the ocean of randomness.

As impressive as these huge concentrations of matter are, they are still, on the whole, rather simple structures—balls of hydrogen and helium held together, and collected in groups, by forces of gravity of such magnitude as to be beyond conception for the unpracticed mind. Thus even a large atom is a rarity, and the conglomeration of protons and neutrons present in, for example, an iron atom represents a minute but deep pocket of negative entropy. Atoms often bond together, and thus hydrogen often exists as H_2—two linked atoms. When atoms are bonded a higher order of organization exists, but this level of complexity is carried another step further when two or more different elements become linked. Water, a molecule composed of two hydrogen atoms and an oxygen atom, is an extraordinarily rare finding in the cosmos, and its presence in a liquid state is considered to be an important, if not essential, requirement for the evolution of life. Because random dispersal of matter is favored by the law of entropy, energy is required to create and maintain organized structures in the form of bonds between atoms.

Planetary bodies frequently contain molecules composed of atoms of different types. Methane (one carbon, four hydrogens) and ammonia (one nitrogen, three hydrogens) are readily detected on several of our own solar system's planets. However, planetary bodies are exceedingly rare in the known universe. Despite intense searching with powerful telescopes for many decades, it is only within the last few years that we have been able to establish even the existence of planets outside our own solar system. Thus even simple molecules are extremely rare finds.

It is in this context that we should contemplate the molecules found in living organisms. Some of these molecules contain not four or five, but tens of millions of atoms! These molecules contain many different kinds of atoms as well. Not only are these extraordinarily rare, complex aggregations of atoms found in living organisms, but even the simplest such organisms contain several different kinds of these "macro" molecules. As such, living entities are by far the deepest depressions in the surface of the entropic sea. They are narrow but incredibly deep vortexes. As an extreme violation of the tendency to randomness, it should not be surprising that the living state is very unstable and lasts, in cosmic time, only a fleeting moment. A human being lives about 100 years. If the life of an average star like the sun was measured as 100 years, a human life, by extrapolation, would come and go in little over 3 seconds.

The characteristics of living matter that allow us to distinguish it from the inanimate emerge from the tremendously highly organized state of the material within it. It is not the matter per se that makes something alive, it is the state of organization of that matter. To appreciate this point, it is worthwhile to consider the fact that nearly all of the atoms in a human being are constantly turning over: Atoms leave the body and are replaced by new atoms of the same type. Even bone is not static but is chewed up at its interior surface as new bone is laid down at the exterior. Thus the calcium atoms present in bone at any given moment are just passing through, and it is not the calcium that makes bone what it is, rather, it is the way the calcium atoms are organized. From this perspective the essence of a human being may be regarded not as its solid structure but as an arrangement of matter. Much like the rainbow on the waterfall, as Thomas Mann would put it, it is the arrangement of the material that allows it to behave in a manner that we recognize as animate. Without seeming excessively deprecatory, a human life may be regarded as analogous to that vortex we see in the water when a toilet is flushed: Water molecules pass through this structure, which requires energy to create (the force of gravity) and which quickly dissipates to restore a condition of randomness.

The foregoing discussion should make clear the point that to generate large molecules composed of many kinds of atoms, as occurs in the macromolecules of living organisms, energy is required. But energy is not enough. Conditions must also be created that allow atoms to come into close proximity so that bonds between them can be formed. A key factor in the evolution of life was almost certainly liquid water. As rains fell on the primeval, very hot earth, the hot water dissolved elements in the rock and allowed them to interact closely in a liquid medium. As early pools of water evaporated, these elements became very concentrated. With the energy of the earth's heat and the electrical energy of storms, bonds between different atoms were able to form, generating the early subunits of macromolecules. In fact, it has been possible to place nitrogen, carbon, hydrogen, and oxygen into a vessel in the laboratory, apply a jolt of energy, and create an amino acid—a complex molecule that constitutes the basic subunit of protein.

A remarkable finding made by the Russian scientist Oparin in the early part of the twentieth century provides insight into what might have been the next step in generating a living organism from inanimate parts. When Oparin assembled a variety of molecules that are frequently encountered as subunits of macromolecules found in living systems, and mixed these compounds with water, so-called "coacervate droplets" formed. These droplets have the remarkable feature that they allow for at least a transiently stable association of different molecules, in resistance to the tendency toward randomness. The stability of coacervate droplets arises from the fact that in the microcosmic environment of these molecules mixed in water the association of these molecules cannot be disrupted without a small amount of energy input. Eventually, of course, the components of these droplets will become disassociated; but, as already alluded to, a living being is also a transitory state of highly organized matter.

How do we take the step from a transiently stable association of molecules that form the subunits of the macromolecules of living organisms to the creation of a living entity? It's easy enough to imagine further bonding of these molecules to form even larger molecules. But what we must envision as a key step in the evolution of life is a point at which large molecules were able to utilize small quanta of energy, perhaps from the environment, or perhaps released when bonds in neighboring molecules were broken, to replace atoms lost through damage. That is, conditions would have to arise under which large molecules could be so predisposed to maintenance of a transiently organized state that they are able to utilize energy and available elements to repair themselves. Although the energy to effect such repair could be gleaned from any source, the harvest of energy from broken bonds of other molecules is a particularly attractive source, because the availability of that energy is not dependent on chance. It should be appreciated that because energy is neither created nor destroyed (the first law of thermodynamics), bonds created through energy expenditure must perforce release energy when they are broken. Thus it is possible to store energy in the form of bonds in expendable molecules.

From this point, we might consider evolution of conditions under which molecules are not only capable of repair, but also replication. Replication and repair are actually closely related processes. If a macromolecule loses a subunit, repair entails replacement of that subunit. Similarly, the order of subunits in a macromolecule might be used as an information "template" for de novo creation of a new identical macromolecule by assembly of the corresponding subunits. Although replication is not necessarily an essential feature of a living entity, it is not surprising that evolution would extend the tendency to engage in repair to its most effective form—replication.

Once molecules have characteristics that, in the appropriate environment, allow them to replicate, a major step toward the evolution of a living state is accomplished. These highly organized, but perhaps not yet "living," states are perhaps what Mann was envisioning as representing the "transition between vitalized organization and mere chemistry."

The creation of a free-living organism requires the ability not only to replicate key macromolecules but to expend energy in such a way as to assemble the resources needed for replication. Thus, whatever molecules are needed to effect repair or replication, information for their production must be contained within the structure of the "master" macromolecule. Most viruses are not free living. Although they are able to replicate their macromolecules when provided with the appropriate building blocks and other molecular tools, they are required to "borrow" one or more of these components from another living entity. The smallest free-living entity is a cell, which is able to create a partition between itself and its immediate environment, sequester the resources needed for replication of its master macromolecules, and produce the molecules needed to carry out the replication process. Mammalian cells, which are far more complex than bacterial cells, have many subcompartments within them, and within these each of these smaller compartments a subset of the tasks needed to maintain the living state and to replicate is carried out. In a real sense, a mammalian cell is a miniature organism, with its subcompartments, or organelles, carrying out tasks not dissimilar from those assigned to the organs of the body of an adult. Figure 1 shows a cartoon of a cell with the key subcompartments shown in the context of the functions they carry out.

An adult human is a colony of cells. Over the 30 million centuries or so that evolution has been at work, cells in these colonies have acquired specialized functions that allow them to maintain the colony as a whole. Of course, all cells in the colony must be able to keep themselves alive, and thus some of the maintenance, or "housekeeping," information is the same in all of them. However, the specialized functions that cells in specific organs must perform to keep the colony going are not performed by other cell types. Theoretically, the information required to direct these specialized activities need be present in only a small subset of cells in the colony. However, as we shall see, evolution has taken the conservative approach of providing all cells in the colony with all of

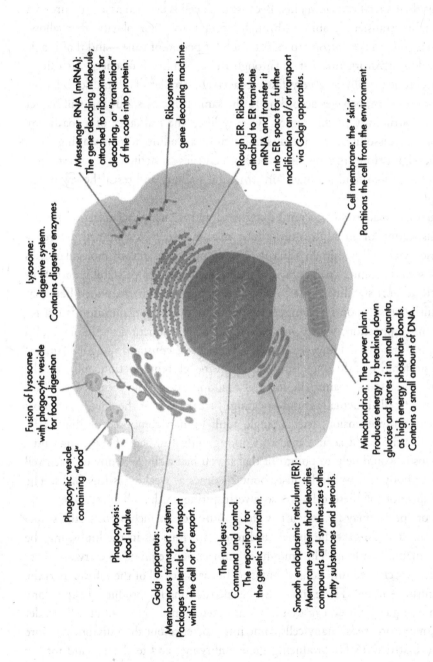

Messenger RNA (mRNA): The gene decoding molecule, attached to ribosomes for decoding, or "translation" of the code into protein.

Ribosomes: gene decoding machine.

Rough ER. Ribosomes attached to ER translate mRNA and transfer it into ER space for further modification and/or transport via Golgi apparatus.

Cell membrane: the "skin". Partitions the cell from the environment.

lysosome: digestive system. Contains digestive enzymes

Fusion of lysosome with phagocytic vesicle for food digestion

Phagocytic vesicle containing "food"

Phagocytosis: food intake

Golgi apparatus: Membranous transport system. Packages materials for transport within the cell or for export.

The nucleus: Command and control. The repository for the genetic information.

Smooth endoplasmic reticulum (ER): Membrane system that detoxifies compounds and synthesizes other fatty substances and steroids.

Mitochondrion: The power plant. Produces energy by breaking down glucose and stores it in small quanta, as high energy phosphate bonds. Contains a small amount of DNA.

Figure 1   A simplified cartoon of a single human cell.

the information required to make and maintain the adult organism and has taught cells how to utilize only the information they need to stay alive and carry out their specific assignments. Because each cell is evolved as a structure that maintains transient stability through complexity—complexity that allows pockets of negative entropy to exist for a brief period of time—and that is able to prolong this organized state through the expenditure of energy, it can be viewed conceptually as a very fancy coacervate droplet. The multicellular organism, in turn, behaves according to the same principles, but at a still higher order of structure. In this regard, we may liken the resistance of a coacervate droplet to dissolution as not dissimilar from an adult human waiting for a "walk" sign before crossing the street: Both entities are acting, at least transiently, in a way that tends to maintain structural stability and resist the law of entropy.

Such an analogy should not be confused with a lack of appreciation for the sophistication of an adult multicellular mammal such as a human or even a mouse. Each cell within an adult is carrying out thousands of molecular tasks every second, some of which involve maintenance of its own stability, others of which support stability of the colony. It is important to be reminded here that the ultimate solution to maintenance of a state of high organization is not repair but replication, which we understand as reproduction at the level of a complex organism. However, repair processes are constantly ongoing, maintaining stability of the organism until it is, through reproduction, able to make its strongest stand against the force of randomization. The complexity of this colony of cells is actually mind boggling. Just the size of the colony is almost beyond comprehension, with a single adult human composed of about 10^{14} cells. For those not accustomed to working with numbers of this magnitude, perhaps it would help to point out that if you had a single penny for each cell in your body you would be just about 20 times as wealthy as Bill Gates. The specialization of these cells is also awe inspiring. Although some, like blood cells or sperm, may be only a few thousandths of an inch across, others, like neurons that connect the spinal cord to distant muscles in the limbs, may be more than a foot long! The amount of information required to carry out these diverse, specialized functions is fantastic, but it is not all of the information the organism requires. There must also be instructions for producing the colony from a single fertilized egg that bears no resemblance to any adult cell. As development proceeds, many cells come into existence but then disappear before birth. Instructions for producing these embryonic and fetal cells, and for the specialized functions carried out by these cells, must also be present in the information data bank. How does the organism store this incredible amount of information, maintain its quality and integrity, and access the portions required at the right time? The method is creation of molecules that have this information as an intrinsic part of their structures. The storage and access procedures are both fascinating and critical to an understanding of the potentials and limitations of genetic engineering. We will therefore examine them in the next chapter.

3

MOLECULAR BIOLOGY AND RECOMBINANT DNA TECHNOLOGY

It has not escaped our notice that the specific pairing we have postulated immediately suggests a possible copying mechanism for the genetic material.

—James D. Watson and Francis Crick, 1953

The Structure and Function of Genes, the Informational Macromolecules

From the foregoing discussion it should not be surprising that the large macro-molecules that contain the information needed to produce and maintain a multicellular mammal like a human are composed of subunits organized in groups: It is far easier to store complex information as arrays of readily available smaller molecules than it is to create an enormously complex molecule de novo. The key macromolecule is of course that from which genes are made, de-oxyribonucleic acid, more affectionately known as DNA.

Incredibly, DNA is made from only four different kinds of subunits, ade-nine, guanine, cytosine, and thymine, which for now we will simply call A, G, C, and T. How can all of the instructions needed to produce and operate such a fantastically complex structure as a human being be provided with only a four-letter alphabet? A useful way of understanding the approach to this problem is to think of the way in which we write large numbers with small number systems. In the binary number system (base 2) there exist only two numbers, 0 and 1. If we want to symbolize the number 2 in this system we must write 10. The number 4 in the binary system is 100, and the number 32 is 100,000. There is no number too large to be expressed in the binary system, but it is easy to appreciate that in this number system with only two digits we must exploit length to represent the larger numbers.

The genetic material, DNA, has only 4 subunits, which is more than the 2 numbers of the binary system but far fewer than the 9 digits of the decimal system or the 26 letters of the English alphabet. Therefore, if DNA is going to write the instructions to produce a brain with tens of millions of interconnected neurons, a heart that pumps 100,000 times per day for up to 100 years without failing, an immune system with cells that can recognize thousands of foreign invaders and kill them without accidentally attacking their own fellow colony members, a reproductive system that will initiate and support the development of a baby, and the myriad other specialized structures and organ systems of a human, it is going to have to rely on length.

How long is the DNA molecule? The number of subunits, referred to in biology jargon as bases, required to encode the development and function of a human is about three billion. If the genetic code were written out at 12 characters per inch, a size typical for a printout from a word processor, the total length of the printout would be about 47,000 miles. Of course, the characters—the A, T, G, and C bases, are submicroscopic in size. However, even though the DNA bases are exceedingly small, the actual length of DNA required to encode a human is still about one meter. This meter of DNA is packaged into the nucleus of the cell, a structure only a few millionths of a meter across. Consider for a moment the efficiency of this information packaging system: A typical CD-ROM disk holds about 650 megabytes of information and is about 6 inches across. The nucleus of a cell holds 6 gigabytes of genetic information (each nucleus actually has two complete copies of the 3 billion-base genetic code), in a structure far too small for the naked eye to see. CD-ROM drives or even computer hard drives, which can hold a bit more than 6 gigabytes of data, are fairly durable—we expect them to last a year or two at least. But the nucleus of a cell functions essentially flawlessly for up to 100 years or more! Moreover, over the course of this time, the genetic information data bank is being selectively utilized, thousands of times per second, without an error. If you have any respect at all for that PC on your desk, you should bow with humility and reverence to your nuclei.

The four bases of DNA are linked in essentially one chain of three billion subunits, and, as alluded to above, there are two such chains in each cell nucleus. Understanding the way in which they are linked and associated to produce the functional DNA molecule is key to understanding genetics, development, and recombinant DNA technology, all of which are disciplines important to the design and execution of genetic manipulation strategies. Fortunately, the fundamentals of DNA structure are quite simple.

Figure 2 shows the four DNA bases in cartoon form. The two chains in the DNA are solidly formed with a "lock and bolt" bonding. In this figure the locks are left slightly open to indicate the actual site of atomic bonding, which will be shown later. The ring bolt cannot be opened, so when a chain of bases is made by joining the locks to the ring bolts, this joining process can occur in only one direction. The two chains of locks are attached to the bases of DNA, indicated by the letters A, T, G, and C. Note that the A, T, G, and C bases are associated in the two chains by attractive forces, indicated by the red "lightning bolts" between them. These forces are not as solid as the lock mechanism on the outside of the chain but are attractive forces more akin to magnetic forces. These forces are referred to as hydrogen bonds. These hydrogen bonds are in reality pretty weak. The DNA strands are readily melted apart by heating, and they will also separate in alkaline solutions. The melting and reannealing of hydrogen bonds by heating and slow cooling can lead to production of free-floating single strands that, when given time through slow cooling, will find the partner strands and reform the double helix in a process referred to as reannealing. It is truly remarkable how perfect reannealing is when one considers the

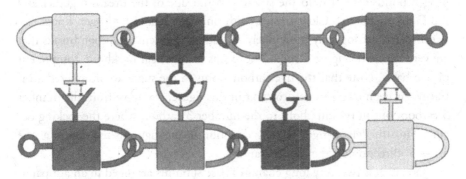

Figure 2   Two strands of DNA, joined as locks, and associated (A with T, or red and yellow, and G with C, or purple and green) via hydrogen bond attraction. There are three points of attraction between G and C and two between A and C.

large number of bases in the DNA. Two other features of the A:T and G:C associations are important to notice. One is that this association is based on the shapes of the bases. Thus G can only pair with C and A can only pair with T, because any other base association will not "fit." A second feature of these attractive forces is that there are three points of attraction between G and C and only two between A and T (Figure 2). Also note that the locks, although shown in different colors for each base, are the same in structure for all for bases. Finally, in the two DNA strands the locks are positioned in an antiparallel fashion, such that they run from left to right in the lower strand and right to left in the upper strand. The sugars, shown as locks with a ring bolt, are not perfectly symmetrical structures but rather have a rear end (the ring bolt) and a front end (the lock). In Figure 3 these bases are shown attached to the backbone of the DNA molecule, the deoxyribose sugar. The "locks" represent the sugar-phosphate portion of the subunits. Note that now at the molecular level this portion is the same in all four bases. The unique part of each base, which distinguishes it structurally and functionally from the other bases, is depicted in the cartoon as simply the letter A, T, G, or C superimposed in the molecular structures of the bases. The A, T, G, and C components protrude from the sugar backbone.

At body temperature, the billions of hydrogen bonds between the strands keep them stable but also are easily separated in localized regions when enzymes need to access a strand for DNA replication or gene expression (see below). It should be apparent that if two breaks occur in the sugar backbone on opposite strands, the hydrogen bonds between the breaks are all that is left to hold the strands together. Thus, when breaks occur in the backbone, the hydrogen bonds cannot hold the strands on the side of the breaks together and the DNA fragments. Close breaks, with only few sets of hydrogen bonds between them, are logically more likely to lead to fragmentation than breaks that are very far apart. Figure 4 shows the atoms in the sugar backbone component of two bases. Note that the five carbon atoms in the sugar are numbered arbitrarily. As we move from left to right in this figure, we move from the number 5 carbon, out in the ring bolt, to the number 3 carbon, where the locking occurs. Thus the left-to-right reading direction in this figure is referred to as the $5' \rightarrow 3'$ direction on the DNA strand.

DNA is thus two very long chains of base subunits arranged in an antiparallel configuration and associated with each other through the additive forces of the relatively weak attractions between the A:T and G:C pairs. In the nucleus these strands are twisted around one another in a spiral arrangement, thus ex-

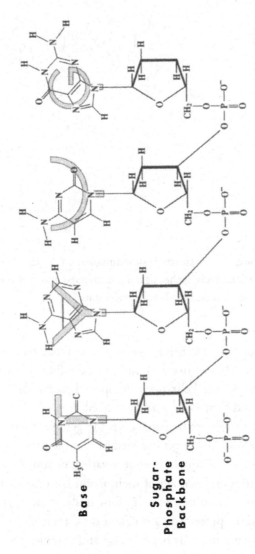

Figure 3   The A, T, G, and C molecular structures of the bases are shown with the letters superimposed. They are shown attached to the sugar (deoxyribose) phosphate backbone, this portion of which is the same for all attached bases.

25

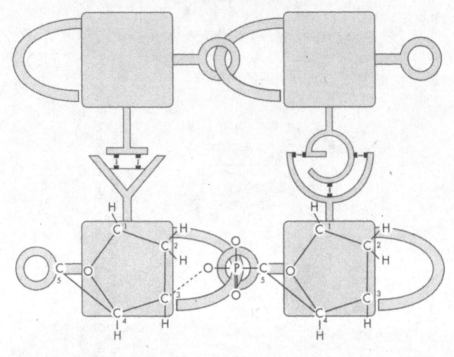

Figure 4   The deoxyribose atomic structure superimposed on the lock components of A and C. Notice that the 5 carbons of the sugar are numbered and that the numbering provides a 5 → 3 directional orientation for the backbone.

plaining the description "double helix." To copy the DNA strands when the DNA replicates, or use the bases in the strands for decoding instructions to direct cell functions, the double helix must be opened so the A, T, C, and G bases can be accessed and their order read by the cell.

The four bases are the letters that make up the genetic alphabet, and the words are genes. How does a gene get decoded so that the function it specifies can be carried out? Genes, of course, encode a variety of functions, but a fair and reasonably accurate simplification of their action is to say that each gene carries the code for production of a protein. Obviously, there are many thousands of different kinds of proteins in a cell, and these must all be encoded by genes. Because these many proteins have diverse and complex functions, they must also be macromolecules, and the logic of evolution again predicts that these complex macromolecules will be assembled as chains of subunits. The subunits of proteins are amino acids, whose structure will be examined momentarily. Because there are only four different bases in DNA, a "one base:one amino acid" correspondence would result in the existence of only four different amino acids. Such a limited diversity of amino acids would make it quite difficult to generate the many thousands of different proteins needed to carry out both housekeeping and specialized functions in the cell. Therefore, to increase diversity the genetic code is organized as three bases for every amino acid. Table 1 shows that use of a three base:one amino acid code generates more than twenty different possible codes. Note that four of the triplet codes are highlighted. The AUG triplet in the first column encodes the amino acid methionine (met), and the special importance of this triplet will be discussed later. The highlighted triplets in columns 3 and 4, UAA, UAG, and UGA, do not encode amino acids but perform other functions that again will be described shortly. It is easy to see from this table that some amino acids have more than one possible code. However, although the cell may use more than one triplet

Table 1. Genetic code

		Second Base			
		U	C	A	G	
First Base (5' end)	U	UUU UUC Phe / UUA UUG Leu	UCU UCC UCA UCC Ser	UAU UAC Tyr / UAA UAG	UGU UGU Cys / UGA UGG Trp	U C A G
	C	CUU CUC CUA CUG Leu	CCU CCC CCA CCU Pro	CAU CAU Tyr / CAA CAU Tyr	CGU CGC CGA CGU Arg	U C A G
	A	AUU AUC Ile AUA / AUG	ACU ACC ACA Thr / ACG	AAU AAC Asn / AAA AAG Lys	AGU AGC Ser / AGA AGG Arg	U C A G
	G	GUU GUC GUA Val / GUG	GCU GCC GCA Ala / GCG	GAU GAC Tyr / GAA GAG Glu	GGU GGC GGA Gly / GGG	U C A G	Third Base (3' end)

code to produce an amino acid, it never makes the mistake of using one code for any more than one corresponding amino acid.

It is not an essential part of this discussion to examine in detail the structure of amino acids, although they are presented in Figure 5 for reference. Like DNA bases, each has a "back end" and a front end. The back end has a nitro-

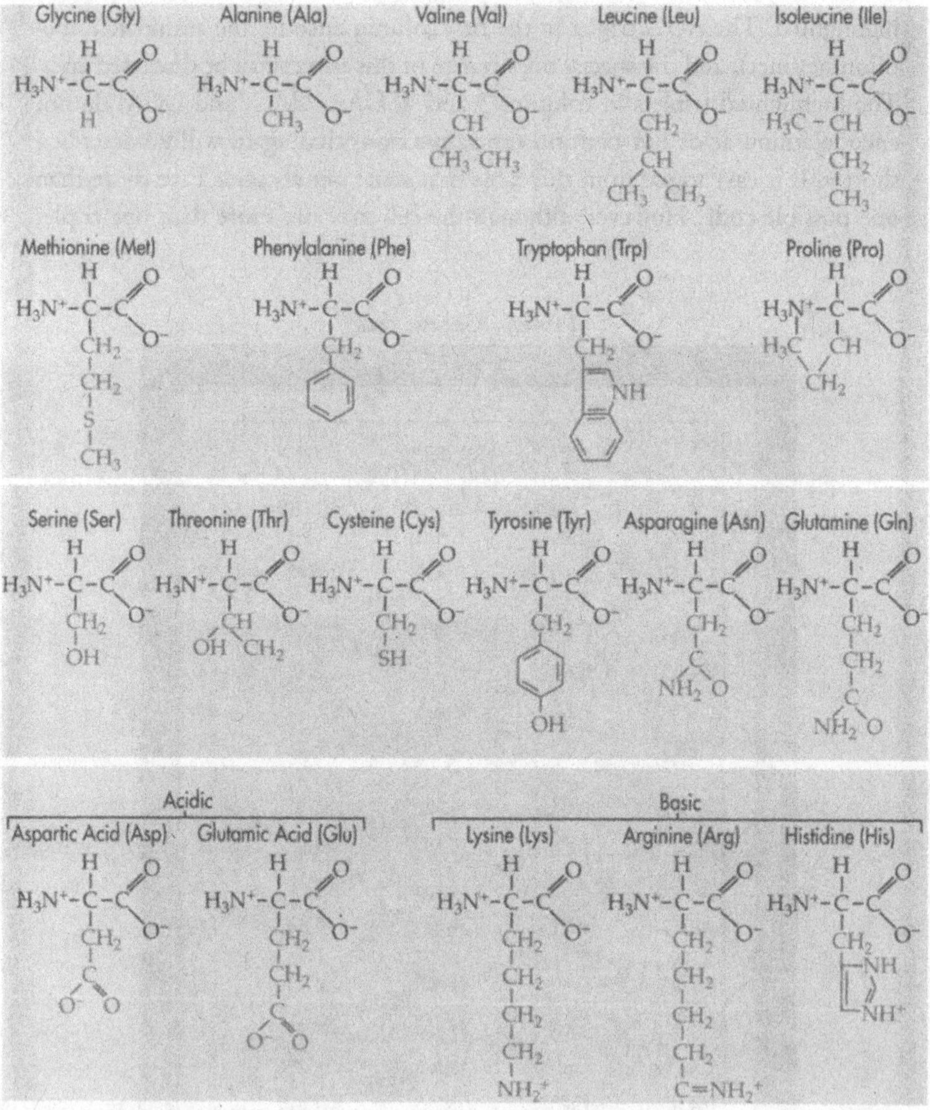

Figure 5 Structures of the 20 amino acids found in proteins. The different side chains are highlighted in red.

gen and two hydrogen atoms (in the figure the nitrogen is shown with three hydrogen atoms and has a positive charge), and the front end has a carbon atom, two oxygen atoms and a hydrogen atom. The diversity of amino acid structure is achieved by linking side chains of atoms to the main "chassis" of the amino acid. The side chains in Figure 5 are shown in red. Some of these side chains are very simple—just a hydrogen atom in the amino acid glycine. However, others are more complicated, with more atoms and a greater variety of atoms. For example, the amino acid cysteine has a sulfur atom in its side chain. When strings of amino acids are put together to make a protein, they fold in very specific ways, and this three-dimensional folded structure confers upon the protein its specialized function. Figure 6 shows four amino acids linked as they are in a typical protein. Notice that the start of the protein at the left has the NH_2 group, and the end has the COOH group. The NH_2 end is referred to as the amino terminus and corresponds to the 5' end of the DNA sequence, whereas the COOH group, or carboxy terminus, corresponds to the 3' end of the gene (again, 5 and 3 refer to the numbers of the carbons in the deoxyribose sugar). Proteins may be structural like collagen, may be locomotive like the proteins of muscle, may be hormones like insulin, or may be enzymes. Enzymes assist cells in carrying out the chemical reactions needed for life by making those reactions more efficient. Enzymes are named for their function

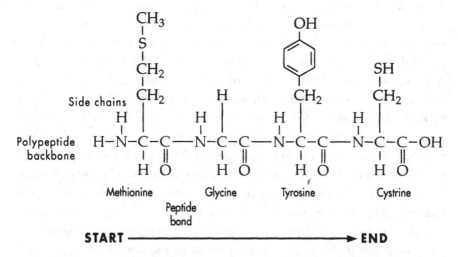

Figure 6   Three amino acids linked in a chain as is found in proteins. Note that the NH2 group of the first amino acid marks the starting point of the protein and the carboxy group (COOH) is at the end.

and then have the suffix "ase" to indicate that they are enzymes. Thus proteases are enzymes that digest proteins, DNA-ase digests DNA, DNA polymerase makes a new DNA polymer and thus synthesizes new DNA, and so on. For the purpose of this book we will consider only a few proteins, chosen because they are exemplary of specific developmental functions.

Copying the DNA

If the cell is going to reproduce, it must copy its genetic information, the DNA. Copying is actually a rather simple process, assisted by enzymes, which are of course proteins that are encoded in the DNA. When the cell has mustered the resources needed to copy the genetic material, resources that include 6 billion pairs of bases and molecules to provide the energy needed for formation of new bonds between the sugars on the new strands, it "pries" open the double helix and uses the enzyme DNA polymerase to scan the bases. Then, wherever a T is present on the strand to be copied, the enzyme inserts an A, and wherever a G exists, a C is placed opposite. Note that the cell does not have to "think" to know which base to place opposite each exposed base in the old strand, because only one base will fit. Therefore, insertion of the bases in the correct order is more or less automatic. The beauty of this system was recognized by James Watson and Francis Crick, who originally deduced the structure of DNA and who were thus inspired to produce the statement quoted at the head of this chapter. Another key point to appreciate is that the new strands cannot both be produced by exactly the same linkage strategy, because linkages can only be formed in one direction (to review this point, see the open locks and closed bolts in Figure 2). This direction is $5' \rightarrow 3'$ with respect to production of the new strand. Recall that the $5' \rightarrow 3'$ direction refers to the arbitrary numbers assigned to the carbons in the sugar backbone (see Figure 4). When a base is fixed in position on the new strand and the open lock is exposed, it is possible to slip the ring bolt around the lock. However, when a base is fixed in position and the ring bolt is exposed, it is not possible to slip the lock portion of the next base around it. Therefore, although one strand can be smoothly synthesized by slipping the ring bolt into the open lock and the closing the linkage, this cannot be done for the opposite, or antiparallel, strand, where the exposed part of the sugar is the closed ring bolt. To get around this problem, the strand oriented in the unfavorable direction is copied by making short segments of new DNA in the only possible direction and then linking the segments to one another to complete the process. This special linkage process must use a different enzyme than DNA polymerase, which can only slip bolts over locks. This other enzyme is called DNA ligase, and its ability to join breaks in the DNA backbone make it very useful for recombinant DNA procedures, as we shall see later.

This difference in the copying of the strands is shown in Figure 7. It is remarkable that over the billions of years of evolution, no DNA polymerase has been found that will form the bonds by linking a base into an exposed "bolt" component of the sugar. It would seem far more efficient to have smooth strand extension in both directions, but evolution has been very conservative on this point. The fact that strands can be extended in only one direction will have relevance to our later discussions of genetic engineering. Thus new DNA strands are made in the 5′ → 3′ direction.

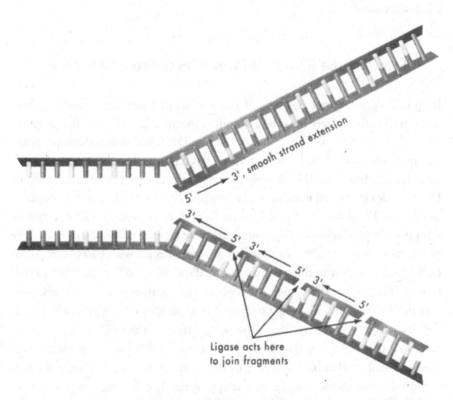

Figure 7   A diagram of DNA replication. New strands are always made in the 5 → 3 direction, so one strand is copied smoothly (upper strand) while the other is made in pieces that are subsequently linked together (bottom strand).

DNA replication is remarkable for its speed. When a cell is dividing rapidly, the division, or reproduction process, requires about a day. Nine hours of this time is devoted to DNA copying in a phase of the cell cycle known as the S phase (for DNA synthesis). In this nine-hour period, 6 billion pairs of bases are inserted into the new DNA strands. This is about 185,000 bases per second! Because each insertion requires bond formation and thus the expenditure of energy, DNA replication is a major undertaking indeed. It should not be surprising that, despite the fact that base configuration relieves the cell of the task of selecting the correct base to place opposite each corresponding base when a new strand is synthesized, mistakes occasionally creep into the system. For a variety of reasons these mistakes are not normally a major problem, but we will revisit this reality when we discuss genetic manipulation strategies, which could be significantly affected by DNA copying errors that occur as the single-celled fertilized egg reproduces itself to produce the 100,000,000,000,000 cells of an adult individual.

Decoding the Genetic Information to Produce Proteins

If the cell is going to make a chain of amino acids that are each encoded by three bases in the DNA, there must be a decoding mechanism. That is, there must be a decoding machine that scans the bases of the DNA and inserts the appropriate amino acid for each three bases read. Although it is theoretically possible for such a machine to read the DNA sequence directly, the cell uses an indirect method. DNA resides in the cell nucleus, and many proteins are needed in the cytoplasm (see Figure 1). Moreover, the cell wishes to protect its precious DNA, exposing it for protein production for the minimum time necessary. The solution to these problems is to produce an intermediate macromolecule with bases very similar to the DNA, send it to the cytoplasm, and allow the decoding machine to read it there. The intermediate molecule is ribonucleic acid, or RNA, and it is indeed very similar to DNA. The sugar portion (corresponding to the lock in Figure 2) is ribose rather than deoxyribose. The base portion of the RNA is nearly identical to that of the DNA, with one small difference: the RNA employs a base called uracil instead of thymine. Because this RNA carries the coded message for protein production to the cytoplasm, it is appropriately called messenger RNA, or mRNA for short. Finally, mRNA is made only from the coding strand of the DNA and is therefore single stranded. These key differences between DNA and RNA are summarized in Figure 8.

Difference Between DNA and RNA

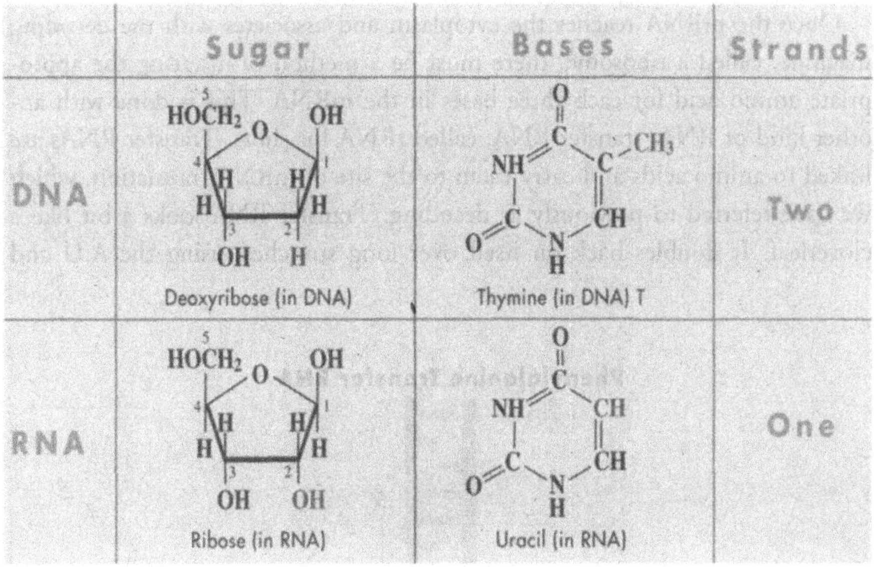

Sugar	Bases	Strands
DNA Deoxyribose (in DNA)	Thymine (in DNA) T	Two
RNA Ribose (in RNA)	Uracil (in RNA)	One

Figure 8 The differences between DNA and RNA. The molecular differences between the sugars and bases in these two molecules.

It's quite easy to make mRNA from the coding strand of the DNA by using a base-pairing mechanism not at all dissimilar from that used for DNA replication. The bases are configured such that when RNA is made the only base that will "fit" is opposite the corresponding DNA base. When the DNA base is adenine, the only RNA base that can pair with the adenine and be placed opposite during RNA synthesis is uracil. Of course, the structural restrictions for G:C pairing are no different when mRNA is made than when DNA is replicated. Again, the cell doesn't have to "think" to put in the correct mRNA bases, because only the right one will fit opposite the corresponding DNA base. The RNA is produced from DNA with an enzyme, of course, and as you might predict, the enzyme is called RNA polymerase. As with DNA replication, the reading, or "transcription" of mRNA can only proceed in one direction; again, it is the 5' → 3' direction with respect to the orientation of the newly synthesized RNA strand. However, in the case of mRNA production, strand extension is not a problem, because only one strand of the DNA is used for mRNA production and this is the one that allows smooth strand extension. It follows that, unlike the double-stranded DNA molecule, RNA molecules, which are produced by using only one of the DNA strands as a template, are single stranded.

Once the mRNA reaches the cytoplasm and associates with the decoding machine, called a ribosome, there must be a method of inserting the appropriate amino acid for each three bases in the mRNA. This is done with another kind of RNA, transfer RNA, called tRNA for short. Transfer RNAs are linked to amino acids and carry them to the site of mRNA translation, which we have referred to previously as decoding. Transfer RNA looks a bit like a cloverleaf. It doubles back on itself over long stretches, using the A:U and

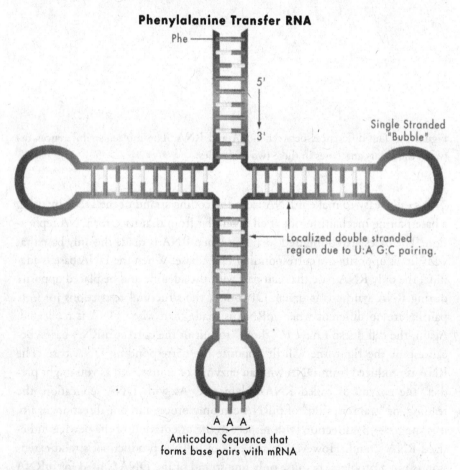

Figure 9 Phenylalanine transfer RNA. Note that some regions are double stranded and that the three adenine residues at the bottom align the molecule with the mRNA for insertion of the attached amino acid.

G:C pairing within different parts of the strand to form localized areas of double-stranded RNA. Bulging from one of these double-stranded regions is a single-stranded "bubble" with three bases exposed. When these three bases are all adenine as shown, the tRNA carries with it the amino acid phenylalanine. The phenylalanine tRNA approaches the ribosome, and its three exposed As form hydrogen bonds with three exposed U bases on the mRNA (remember the specificity of G:C and A:U pairing). This allows the tRNA to become closely associated with the elongating protein strand, and its phenylalanine is then linked to the protein. This mRNA translation strategy underscores the importance of the specificity of base pairing to the cell. DNA is replicated by exploiting it, mRNA is produced by exploiting it, and tRNAs deliver their amino acids to the growing proteins again by taking advantage of the specific nature of base pairing. Figure 10 diagrams this sequence of events, with mRNA produced in the nucleus and translated in the cytoplasm. The methionine RNA is annealed to the mRNA.

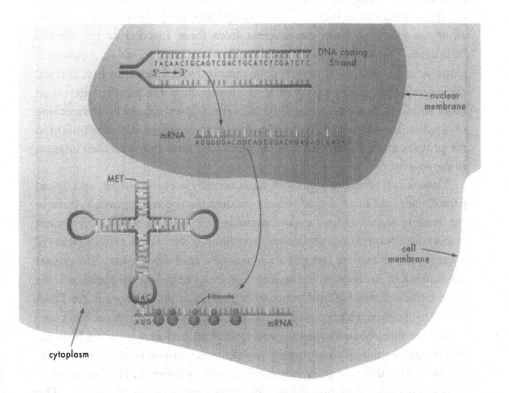

Figure 10 Production of mRNA, transport of mRNA to the cytoplasm, and alignment of tRNA to the mRNA at the ribosome where translation is performed.

Let us now make a simple protein of five amino acids. As you should be able to predict, this protein will require a 15-base DNA code. In our protein we will use the code TAC, AAG, GTT, GCT, CAC. The mRNA will have the sequence AUG, UUC, CAA, CGA, GUG, and the amino acids inserted will be methionine, phenylalanine, glutamine, arginine and valine (see Table 1 for the amino acid code). As shown in Figure 10, the mRNA is made within the nucleus, crosses the nuclear membrane, and associates with the decoding machine, the ribosome, in the cytoplasm. The ribosome then reads the mRNA like a ticker tape, inserting the correct amino acid for each three bases. The process of protein production also involves the action of enzymes, and the linkage of the amino acids, which entails new bond formation, requires energy.

As long as we have this small protein in front of us, let's take a quick look at the effect that base changes in the DNA will have on the final product. For example, let's simply remove the fourth DNA base in this code. If we do this, we now have the sequence AUG, UCC, AAC, GAG, UG. We can see that after the initial methionine is coded, the reading frame of the bases is changed and all of the following bases are different from those encoded by the original "minigene." Now the order of amino acids is methionine, serine, asparagine, glutamic acid, and. . . . However, if we remove DNA bases four, five, and six, the single amino acid phenylalanine is missing from the protein whereas all other amino acids are unchanged. Perhaps the simplest change is the switching of one base for another. This often leads to a single amino acid substitution in the protein, and the importance of such substitutions varies with their location in the protein.

These kinds of changes in DNA bases actually occur in nature, and, as expected, they lead to an altered protein product. As you might predict, shifts in the reading frame, as in the first example, are likely to lead to far greater changes in a protein than the loss of three bases in the correct reading frame or a single base change resulting in a single amino acid substitution. The greater the number of amino acids changed in a protein, the greater the likelihood that the protein will not function properly or perhaps at all. Changes in the DNA bases are of course what we recognize as mutations. Obviously, frame-shift mutations are likely to be devastating and could lead to disease resulting from the absence of function of a key protein. A well-known disease resulting from a single base substitution, or point mutation, is sickle cell disease. This mutation affects the gene for β-globin, the predominant protein of red blood cells. This single amino acid substitution, a valine for a glutamic acid at the sixth position of the globin chain, results from the subsitution of a T for the A in the triplet codon GAG, leading to a GTG triplet. This single amino acid change leads to abnormal protein folding and resultant distortion in the shape of the red blood cell, with all of the disease manifestations so familiar to all of us.

If you fully grasp the principles of gene expression diagrammed in Figure 10, you will also appreciate that mutations in the genes that produce transfer RNA can also lead to insertion of the incorrect amino acid in a protein. For example, if a mutation in the DNA resulted in insertion of an A for the U in the triplet of methionine RNA, the tRNA annealing code would be AAC instead of UAC and would insert a methionine at position where a leucine (encoded in the mRNA as UUG; see Table 1) was intended.

Other Signals in the DNA

The most obvious function of the gene, which is a strand of DNA, is to provide the code for the order of amino acids in the corresponding protein. But there must be other important signals in the DNA strand. The genome contains enough DNA to encode about one million average-sized proteins, but determination of the entire sequence of the human genome, a project recently completed, indicates that there exist "only" about 40,000 genes. Not surprisingly, there is space between genes. But in a DNA strand composed of hundreds of millions of bases, most of which are not within gene coding sequences, how is the cell going to figure out where genes begin and end? Because the only information source available is the sequence of bases in the DNA it follows that there are base sequences that serve as signals, informing the cell where genes start and where they terminate. Surprisingly, these sequences are rather simple. About 30 bases away from the site where mRNA production, or transcription, begins, there is often found a group of A and T bases, the so-called TATA box. Because the gene immediately adjacent to the TATA box will be read in the 5' → 3' direction, with the first base read being on the extreme 5' end of the gene, the TATA box is by convention referred to as "upstream," or on the 5' side of the gene. TATA boxes increase the efficiency with which RNA polymerase associates with the DNA to produce mRNA. TATA boxes are frequently seen, especially near genes that function in only one or a few cell types (for example, globin in red blood cells). In some housekeeping genes, for example, those that help produce new bases for DNA replication and that therefore are expressible in all cells, a TATA box is not found, and instead the 5' flanking region is relatively rich in G:C base pairs. Exactly how G:C-rich regions signal the cell that a gene is nearby is unclear. Another signal that a gene will soon be encountered on the DNA strand is a CAAT sequence, often located about 100 bases on the 5' side of a gene. These CAAT boxes (pronounced like the house pet) are not always present, but again they are frequently seen. These regions of DNA that promote the transcription of their nearby genes are referred to, appropriately, as promoter regions. The region of RNA polymerase binding is referred to as the "promoter region" because signals within it promote mRNA transcription. It is quite surprising how simple these sequences are, given the enormous complexity and size of the genome as a whole.

The first three bases transcribed in mRNA do not encode the first amino acid in the corresponding protein. Instead, there is a flanking region of mRNA that precedes that starting point for translation. How, then, is the decoding machine going to know where to begin translation and to start putting in amino acids? Remarkably, the signal for beginning translation is the same one that encodes the amino acid methionine (AUG in the mRNA). When a methionine triplet code is encountered translation begins, and thus proteins very often begin with methionine. It is for this reason that the AUG triplet in Table 1, column 1 is highlighted. This shockingly simple approach to translation initiation creates another problem, however. If a methionine is needed somewhere in the middle of the protein strand, why does the decoding machine not make the mistake of beginning translation at an AUG sequence in the middle of the mRNA, a mistake that would lead to production of a shortened, defective protein? The cell gets around this problem by placing short but unmistakable sequences of bases just immediately 5' to the AUG that is utilized for initiation of translation. These sequences vary slightly between mRNAs, but an efficient signaling sequence is ACCATGG immediately upstream (5') to the AUG that initiates mRNA translation.

Messenger RNAs do not simply end with the last triplet codon for the last amino acid. There are bases that follow the end of the coding sequence of the protein. These sequences assist in stabilizing the mRNA and stopping translation. If there are sequences beyond the last codon, there must be signals for translation termination. The triplet codes UAA, UAG, and UGA do not code for any amino acid. Instead, they are "stop" codons that tell the translation machine, the ribosome, to stop inserting amino acids. Because of the special functions, these triplets are highlighted in columns 3 and 4 of Table 1. Because these codons stop translation, you should consider the consequences of mutations in the DNA that lead to inadvertent placement of a stop codon in middle of the mRNA rather than the end. Take, for example, the triplet code for the amino acid leucine, UUG. If a point mutation occurred that converted the UUG to a UAG, the ribosome would inadvertently terminate translation and the result would be a truncated protein that would be likely not to function normally. Mutations resulting in premature appearance of stop codons are in fact the underlying cause of a number of genetic diseases.

After the stop codon there is an AAUAAA sextet that helps terminate transcription of the mRNA from the DNA, and then a string of A residues is often added. The "polyA tail," as it is called, facilitates export of mRNA from the nucleus to the cytoplasm and can also stabilize the mRNA. When we say "stabilize," we imply that one mRNA can be used for more than one translation procedure, and this is indeed the case. In fact, an important method for regulating the amount of protein produced from an active gene is variation of the stability of the mRNA.

One more feature of genes and their corresponding mRNA transcripts is important to mention. In bacteria, there is a direct correspondence between the order of bases in the coding strand of the DNA and the order of bases in the translated portion of the mature mRNA. However, this is not the case with mammalian cells. In mammals, the coding sequences of genes are often interrupted by stretches of DNA that are initially transcribed into RNA but quickly spliced out, much like a segment of reel-to-reel tape is spliced. These intervening sequences or "introns," as they are called, are clipped out as the mRNA emerges from the DNA, and the coding regions of the mRNA are then pasted together in the correct reading frame. To splice out introns correctly, there are base sequences in the DNA that signal splicing and assist splicing enzymes in the process of intron removal.

Why would the cell go to all the trouble to produce RNA from DNA sequences that are never translated but simply removed? Recall that insertion of RNA bases requires energy. Some intervening sequences are many thousands of bases long, and thus a great deal of energy is expended to produce them. Yet after they are produced they are simply discarded. What possible reason could the cell have for engaging in such an apparently unproductive activity?

Experiments in which isolated genes are transferred into cells in culture or into animals (we will visit these methodologies below) show that mammalian gene expression is often more efficient when introns are present. Moreover, introns can contain signals that assist the cell in activating tissue-specific genes in the correct pattern. These signal sequences are called enhancers, and they will be discussed a bit more later on. It is important to appreciate, though, that a multicellular organism like a human has a much more challenging gene regulation task than a single-celled organism like a bacterium.

Recombinant DNA Technology

We now have enough knowledge in hand to do some gene cloning, analysis, and splicing by use of recombinant DNA technology. Recombinant DNA techniques are of critical importance to engineering of the human germ line, and thus it is important to have some knowledge of how these procedures are performed. Fortunately, recombinant DNA technology is very simple.

To understand and carry out nearly every technique that would fit within the collective definition of what we call "recombinant DNA technology" you need remember only two things: First, keep in mind that A always pairs with T and G with C, and second, remember that the DNA of all organisms on this planet is identical in structure, such that the A base from any organism, even a bacterium, will pair perfectly well with the T residue in human DNA. With these two basic facts in hand, the rest is easy.

Cloning the Insulin Gene

First, let's clone a gene. To do this, we have to obtain, in relatively pure form, enzymes that will cut, or digest, the sugar backbone of DNA (break the locks; see Fig. 2). For gene cloning we don't want to use a nonspecific DNA-ase (to recall the nomenclature for enzymes, see p. 40) that will indiscriminately destroy the DNA molecule and convert into a soup of free-floating single bases. Rather, we want to use enzymes that cut DNA only when certain base sequences are present. Fortunately, such endonucleases (the prefix "endo" indicating that the cuts are made inside the double strand rather than at the ends), which are restricted in their activity to certain sequences, are readily obtainable from bacteria. Given the characteristics of these special DNA-digesting enzymes, they have the logical name "restriction endonucleases." For brevity and convenience we will refer to these endonucleases simply as restriction enzymes.

Restriction enzymes are made in bacteria as a defense against foreign invaders, typically viruses. They digest viral DNA, fragment it, and thus destroy the virus. Although these enzymes have the potential to destroy the host bacterial DNA as well, they do not run wild like nonspecific nucleases, because they can't cut just anywhere. Thus they are a bit easier to control. Typically, a bacterium will modify the bases that form the restriction enzyme target site just slightly so the restriction enzyme won't work. Thus the bacterial restriction enzyme sites are immune to digestion, whereas those of the invading virus are not.

What kinds of specific DNA sequences are target sites for restriction enzyme digestion? Actually, hundreds of restriction enzymes are known, and although some attack the same sequence there are dozens of different DNA sequences that are targets for attack by different restriction enzymes. Let's take a look at a typical restriction enzyme site. The site we will look at is for the restriction enzyme *Eco*RI. This enzyme, isolated from *Escherichia coli* (thus the prefix *Eco*), a bacterium commonly found in the human digestive tract, digests the following six-base DNA sequence:

5' G•A-A-T-T- C 3'
3' C- T-T-A-A•G 5'

When *Eco*RI sees this sequence, it cuts the sugar backbone between the first and second bases. The position of the cut is denoted with the "•" symbol, instead of a hyphen, which represents the bond between the sugars of the bases. A close look at this sequence will show, however, that it is a mirror image of itself. Thus, if the bottom strand is read from right to left, the same sequence of bases is seen as when the top strand is read from left to right. Because there is no "up" or "down" or "right" or "left" in the cell, *Eco*RI will cut both DNA backbones in the same corresponding place—between the first and second bases when the strand is read in the 5' → 3' direction. Let's now look at the strands after cutting by *Eco*RI:

Left strand: Right strand:
xxxxxxxxxxG   AATTCxxxxxxxxxxxxxxxxxx
xxxxxxxxxxCTTAA   Gxxxxxxxxxxxxxxxxxx

It should be apparent from examining this digested fragment that in this localized region, because only four sets of hydrogen bonds between the central As and Ts hold the strands together, the forces are insufficient and the strands fall away from one another. If you recall that hydrogen bonds are weak, and that the stability of the double helix at body temperature depends on the additive forces of many hydrogen bonds (see p. 23), it should be clear that the eight hydrogen bonds attracting the four pairs of bases that lie between the cuts in the backbone are not enough to hold the strands together. When the strands fall apart, the ends of the fragments have a single-stranded segment that protrudes from the cut end.

*Eco*RI will make this precise cut in DNA isolated from any source, whether it be geranium, anthrax, dog, or human. Now that we have seen how restriction enzymes work, let's see how we can use them to clone a gene. Only one of several workable cloning methods will be shown as an example.

To understand how cloning works, it is important to examine the DNA of a bacterial cell that we will use for producing large quantities of the DNA we want to clone. Figure 11 shows a bacterium with two circular strands of DNA contained within it. One molecule is quite large and attached to the cell membrane, whereas the other is a small, circular molecule that floats freely within the cell. The large molecule contains the genes needed to produce and replicate the bacterium and corresponds to the DNA within the nucleus of a human cell. The small molecule, called a plasmid, has special genes that help the bacterium to survive. It contains genes encoding proteins that destroy antibiotics like penicillin, which, as all of us who have been treated for strep throat know

Bacterial Cell

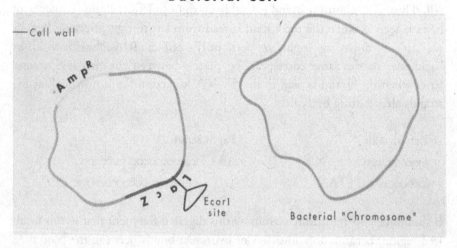

Figure 11   A bacterial cell showing the bacterial "chromosome" attached to the membrane immediately under the cell wall and a smaller circular plasmid that carries genes for ampicillin resistance (Amp^R), and a lacZ gene (see text). Note the presence of a single recognition site for the restriction enzyme *Eco*RI in the lacZ gene.

quite well, is normally very toxic to bacteria. The location of the resistance gene is shown as the yellow portion of the circle and is the gene is labeled Amp^R, for ampicillin resistance (ampicillin is a commonly used form of penicillin). The plasmid contains other DNA sequences that help it to replicate, but this particular plasmid has another gene of special interest, which we will call lacZ. lacZ encodes a bacterial housekeeping enzyme, and it digests certain kinds of sugars. When special forms of these sugars are used, the product of lacZ enzyme reaction is blue. Thus, when bacteria containing a lacZ plasmid are plated on dishes with this sugar included, the bacterial colonies that grow are blue in color. This plasmid has one more important feature relevant to the discussion of cloning: There is a single *Eco*RI restriction enzyme site right in the middle of the lacZ gene. Thus, when this plasmid is digested with *Eco*RI, it opens to give a single linear molecule with part of the lacZ gene on each end. The lacZ gene with its single *Eco*RI site is shown as the blue segment of the plasmid in Figure 11. It should be realized that we can close the circle again with the repair enzyme DNA ligase, which can close breaks in the sugar backbone (see p. 31 for a review of ligase action in DNA replication). Because hydrogen bonds are more stable at lower temperatures, reclosing of the plasmid is accomplished by reducing the temperature to that in a standard refrigerator and then adding DNA ligase. The single-stranded overhangs made by *Eco*RI will tend to reassociate at these temperatures for sufficient time to allow ligase to repair the break made by the restriction enzyme. This repair will close the plasmid circle again and reconstruct the *Eco*RI site.

Through various techniques in the biochemistry lab we can isolate large quantities of such plasmids from bacteria and work with them as chemical reagents. Moreover, plasmids can be introduced into bacteria that don't have plasmids by exposing the bacteria briefly to a high temperature. This heat shock (about a 1-min exposure to a temperature of 42°C) causes bacteria to take up plasmids, after which genes on the plasmids can be activated.

Let us now consider a DNA mixing experiment. We will digest our hypothetical plasmid, which is first isolated from bacteria and purified, with *Eco*RI, thus producing a test tube filled with linear plasmid molecules, each of which has an *Eco*RI overhang at the ends. Presume now that we have a human DNA, and that a gene within this DNA, say the insulin gene, is contained entirely within a fragment that has no sites for *Eco*RI. If we cut human DNA around the insulin gene with this restriction enzyme, we will eventually find an *Eco*RI site upstream from the insulin gene and another downstream. We may have to travel several thousand bases away from the insulin gene in both directions to find these *Eco*RI sites, but the complexity of human DNA ensures that sites will eventually be found. Thus, if we cut human DNA with *Eco*RI, we will obtain a fragment with the insulin gene inside it.

We will now take these human insulin-containing fragments, mix them with the linearized plasmid DNA, and cool the mixture. The staggered *Eco*RI ends will begin to form relatively stable associations of the hydrogen bonds as the temperature is lowered. In some instances, the plasmid ends will find each other and the circular plasmid will reclose. In other cases, the *Eco*RI ends from the upstream and downstream ends of the human DNA will find each other, thus creating little circles of human DNA with the insulin gene inside. But in a third scenario, the *Eco*RI ends of the human DNA will anneal with the *Eco*RI ends of the plasmid. This will happen because of the two principles emphasized at the beginning of this section: A:T and G:C pairing is specific, and all DNA on the planet uses the identical subunits. Because of these facts, human *Eco*RI ends are perfectly capable of forming hydrogen bonds with bacterial plasmid *Eco*RI ends. If this occurs, we can use ligase to create a new circular molecule that is a hybrid between the bacterial plasmid DNA and the human DNA. In this case, the circle will be larger than the original plasmid by an amount equal to the length of the human fragment and the human fragment will be inserted into the middle of the lacZ gene. This ligation product is shown in Figure 12.

We will now add our ligated mixture to bacteria, heat shock the bacteria to make them take up DNA, and then plate the bacteria on an agar plate so they can grow. We are going to use special agar plates containing ampicillin, so bacteria that are not resistant to this antibiotic will fail to grow. We will also add the special substrates for the lacZ gene in the plasmid, so that if this plasmid is in a bacterium and a lacZ gene is expressed, the bacterial colony on the plate will be blue. We will now consider the fates of bacteria that did nor did not successfully take up DNA after heat shock. The possible outcomes of insertion of the ligation mixture into the bacteria, a process called transformation, are shown in Figure 13.

If bacteria fail to take up any DNA they will not grow on our plates because they will not be resistant to ampicillin, which was added to the plates. This is actually what will happen to most of the bacteria, because successful uptake of DNA by transformation is a fairly infrequent event. Bacteria that take up human DNA that has become circularized after ligation will similarly be unable to grow because the human DNA carries the insulin gene but does not carry any genes for penicillin resistance. Bacteria that take up the recircularized plasmid will grow nicely because the Amp^R gene on the plasmid will confer peni-

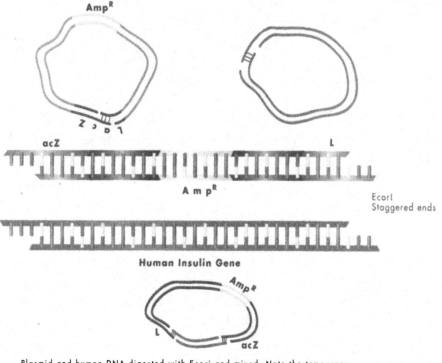

Plasmid and human DNA digested with Ecori and mixed. Note the temporary
association of A:T and G:C bases in Ecori "overhangs", shown in the circular forms

Figure 12

cillin resistance. These colonies will also be blue because reclosing the plasmid
will restore a functional lacZ gene. If a bacterium takes up the new hybrid plasmid, which has the insulin-containing human DNA fragment inserted within
the lacZ gene, it will grow nicely because of the Amp^R gene on the plasmid
DNA. However, this colony will not be blue because the lacZ gene is disrupted
by insertion of the human DNA and cannot function. We will thus obtain a
white colony. This is shown in Figure 13. It's easy to see from this scenario that
identification of plasmids with human DNA inserted into them is trivially
easy—one need only scan plates for white colonies.

Let us now exploit the creation of this new "recombinant" plasmid to produce large quantities of human insulin DNA. We will pick a white colony off
the plate with a clean toothpick and transfer some of the bacterial cells into a
flask that contains broth. This broth contains nutrients that are ideal for

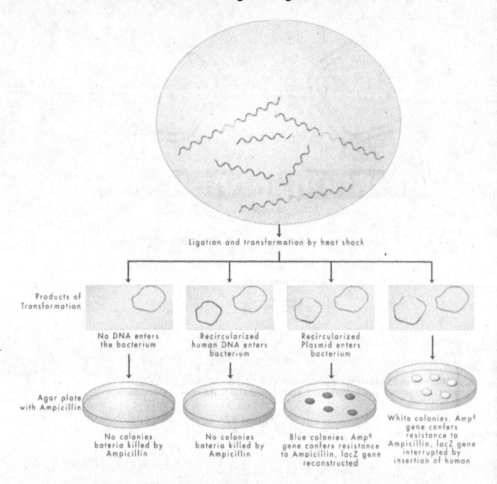

Figure 13   The possible outcomes of exposing bacteria human DNA cut with EcoRI and ligated to a plasmid with an ampicillin resistance gene and a single *Eco*RI site in its lacZ gene. An intact lacZ gene produces a blue colony when plated, whereas a lacZ gene interrupted by a human DNA insertion produces a white colony.

growth of bacteria. Just to make sure we don't accidentally put bacteria that do not carry plasmids into the flask, we'll add penicillin to the broth as well. Any ordinary bacteria that might contaminate the inoculum will thus be unable to grow.

When we grow bacteria in such broth at an optimal temperature for their growth (body temperature, or 37°C), they will divide about once every 20 minutes. How efficient is this growth? Consider a single bacterial cell growing under these conditions. In 20 minutes one cell will divide to become two, and in 40 minutes two will divide to become four. In an hour we will have eight bacterial cells. Eight cells doesn't seem like much after an hour of waiting. But if such a culture were maintained, with more broth added as needed to satisfy the nutritional requirements of the increasing numbers of bacteria, at the end of a month, it would have a greater mass than all of the galaxies in the known universe! In view of our interest in not replacing the entire known universe with bacteria, we can stop the culture after 12 hours or so.

Whenever a bacterium that carries a plasmid replicates its own DNA, the plasmid DNA is replicated as well. In fact, it is not uncommon for the plasmid to replicate many times, and for each bacterium to carry many copies of a plasmid. In our culture system, accidental loss of the plasmid will result in death of the bacterium because loss of the plasmid means loss of antibiotic resistance. Therefore, at the end of overnight culture, our broth will contain tens of billions of bacteria, and with them will be tens of billions of copies of the plasmid. Of course, there will also be tens of billions of copies of a human DNA fragment that contains the insulin gene, because that DNA is replicated with each plasmid replication. We can now purify the plasmids from the bacteria (a simple process that need not be detailed here), digest the plasmids with *Eco*RI, cut our insulin-containing human DNA out of the plasmid, and use it for whatever purpose we wish. One interesting use of such recombinant plasmids is not for isolation of large quantities of human DNA. Rather, we can create conditions that enable the bacteria to express the human gene right from the plasmid. In our present example this could lead to a flask of broth that contained enormous quantities of human insulin. Thus we have cloned the human insulin gene and produced recombinant insulin for treatment of diabetes.

If you are suspicious that this example of insulin cloning is oversimplified, you are correct. A closer look at our DNA mixture after ligation should make apparent the fact that long chains of DNA molecules could be created that would never reclose into a circle. We might get a tube full of spaghetti rather than the nice little circular molecules described in our cloning scenario. Although this is partially true, the reality is that nice circles like the ones we need do form with sufficient frequency that, in technically competent hands, the recombinant plasmids of choice are produced in large quantities and many white bacterial colonies with the correct plasmid in them are easily obtained. There is another more important question, though: Where did we get that *Eco*RI fragment containing the human insulin gene in the first place? This gene encompasses just a few thousand of the 3 billion base pairs in the human genome, and if human genomic DNA is digested with *Eco*RI, millions of fragments will be produced and only a very small percentage of those fragments will carry the insulin gene. Obtaining the human insulin fragment is indeed something of a challenge, and we have ignored the problem. However, finding the insulin fragment among all of those millions of other fragments is labor intensive but not difficult. Let's see how it's done.

First, we digest DNA isolated from human tissue with *Eco*RI; remember that we can get the insulin gene from any tissue, not just the pancreas where the gene is active, because all cells have two complete copies of all of the genetic information, including the insulin gene. We will now mix these millions of *Eco*RI fragments with our plasmid, which has also be digested with *Eco*RI. We will use an excess of plasmid DNA because if recircularized plasmid is introduced into bacteria, they can easily be detected by the fact that the bacterial colony will be blue. On the other hand, we don't want our insulin fragment to form hydrogen bonds with a completely unrelated human *Eco*RI fragment from elsewhere in the genome before being joined in the plasmid. An excess of plasmid makes it unlikely that human fragments will associate with other human fragments rather than with plasmid DNA. We will now add DNA ligase and reconnect all of the *Eco*RI sites. This will be result in a complex mixture of ligation products, but it should be easy to realize that if we use human and plasmid DNA in great enough amounts we will succeed in inserting every *Eco*RI fragment in the human genome into a plasmid. All that remains of our task of isolating a plasmid with the insulin gene inside is to find that plasmid among the millions of others present.

The first step, of course, will be to transform bacteria with the mixture, plate them on plates with antibiotics, and obtain enough white colonies to be sure that one of those colonies contains bacteria with a plasmid that carries the insulin gene. We have to find that colony. To do this we will take advantage of our knowledge of the genetic code and the A:T, G:C base pairing principle.

Insulin is of course a protein, and it can be easily isolated from pancreatic tissue of people who have the tissue surgically removed or who die. With large amounts of the protein in hand, it is relatively simple to determine the amino acid sequence. Once we know the amino acid sequence, we of course know the corresponding sequence of DNA bases that determines the coding sequence of the mRNA. We will now consider a short stretch of this sequence, say 20 bases. We can now use a machine capable of linking short stretches of DNA by using chemical reactions that achieve the same result as DNA polymerase. We simply feed the machine A, T, G, and C bases and plug into a simple computer the order in which we would like these bases to be linked. The machine can then churn out billions of copies of these 20 base fragments that correspond identically to a short stretch of the DNA in the insulin-coding region. It is not necessary for our cloning effort to synthesize both strands of the DNA. One of the strands will do. When we make these short stretches, which we call oligonucleotides, in our machine, which we logically call an oligonucleotide synthesizer, we will add one additional twist. We will make one of the atoms of one of the bases a radioactive isotope, so that the oligonucleotides will be radioactive. This radioactivity is sufficiently intense to expose X-ray film.

We now need to take a small sample of the bacteria in the white colonies without losing track of where they are located on our plates. Each colony has millions of bacteria, so we can easily borrow a few for analysis while saving the rest for growing to obtain high amounts of the insulin plasmid after we find it. To do such sampling efficiently, we can take a piece of filter paper and lay it down on the plates. A few bacteria from each colony will then stick to the filter paper. If we remove the paper carefully, the colonies on the plates will not be disturbed and their positions on the plates will be mirrored on the filter papers. We now have to process these filter papers so that we can search the colonies for the insulin gene. These processing procedures accomplish the following. First, they rupture the bacterial cells, which results in release of the DNA, but in this process, the DNA sticks to the filter in the exact position at which the bacteria were originally situated. We then treat the filters with an alkaline solution so that the two strands of each plasmid molecule separate. Remember, alkaline solutions break the hydrogen bonds (see p. 23). When the filters are dried, the hydrogen bonds are unable to reform. After this processing, which requires only a few hours, we have filters with splotches of single-stranded DNA, positioned in a way that corresponds precisely to the locations of bacteria on our plates. Now comes the key step.

The filters are next placed into a liquid solution that allows single-stranded DNA fragments to form hydrogen bonds with their corresponding opposite strands, of course, by the A:T, G:C principle. Into this solution we will also add our radioactive oligonucleotide, the sequence of which corresponds exactly to one of the strands of the coding sequence of insulin. The oligonucleotides cannot form hydrogen bonds with each other because the synthesizer made only one strand. Of course, the A on one strand will not form hydrogen bonds with the A on another.

However, if we incubate our filters with the oligonucleotides, and we give the oligonucleotides time to "find" a complementary strand of DNA with which to associate, our radioactive fragments will eventually find their partners on the filter. Those partners will be from the insulin gene, which was cloned into a plasmid, placed in bacteria, transferred to the filter from the bacterial colony, rendered single stranded in alkaline solution, and fixed in position. Once our oligonucleotides have found these insulin genes and "hybridized" (formed hydrogen bonds) with them, we can wash the filters to remove excess radioactive DNA, dry the filters, and expose them to X-ray film. Our radioactive oligonucleotide "probe" can of course only hybridize to one of the insulin strands—the one opposite that used to make the oligonucleotide. Once hydrogen bonds have formed between the oligonuclotide and the complementary stretch of DNA from the insulin gene, you should readily appreciate that the DNA from bacterial colonies that contained the insulin gene will now be radioactively labelled. When the filters are exposed the film, those colonies will produce a black dot. From this point it is quite simple to go back to our plates, pick the colony whose DNA hybridized to the oligonucleotide, and grow it in larger quantities so that we can isolate the gene. We have now cloned the insulin gene and established a method for cloning the gene of any protein that we can purify in sufficient quantities to allow determination of the amino acid sequence.

You should also appreciate that this cloning approach actually clones every *Eco*RI fragment in the human genome and thus produces a "library" of DNA fragments. All of the genetic information needed to create a human being is contained somewhere on those bacterial plates, provided we have made enough of them. It should also be clear that our example of insulin cloning was idealized by assuming that the entire gene was contained within one *Eco*RI restriction enzyme fragment. Things would have been a bit more challenging if there were several sites for this enzyme inside the gene. Another problem could arise if there existed a region of the genome in which the *Eco*RI sites were very sparse. If we did not encounter such a site over a one million-base stretch, it would be difficult for a plasmid to carry a fragment of such large size. The fact that plasmids do not carry very large fragments of DNA creates another problem: Cloning human DNA as millions of tiny pieces makes it harder to determine the order in which they are linked together before we cut them apart with a restriction enzyme. The arrangement of DNA sequences relative to one another is very important to know for a variety of reasons.

All of these problems can be readily solved. There exist simple methods of cloning very large fragments (a million bases or more) and of "skipping" some restriction enzyme sites so that genes with multiple sites remain intact. However, the production of a plasmid library such as the one we just created illustrates all of the important principles of gene cloning, so we need not discuss some of the more esoteric strategies.

Amplifying a Fragment of DNA Present in Very Small Quantities to Obtain Amounts Sufficient for Analysis

Gene cloning is an extremely powerful technique, but there are other uses of recombinant DNA technology that are no less powerful. One of these is the ability to take a very small amount of DNA, even a single molecule, and rapidly amplify it so that we have sufficient amounts for analysis and use in experiments. This technology has become critically important for screening embryos for genetic disease. The amplification process is called the polymerase chain reaction, or PCR. As you should be able to predict by now, PCR relies for its success on the A:T, G:C base pairing principle. What else?

PCR is done when the sequence of interest is already known. Consider a DNA fragment of 376 base pairs surrounding the base that is altered in sickle cell anemia (see pp. 36–37). This segment of DNA sits within a much larger molecule, of course, but for the moment, we will consider only this small region. We will now use the same oligonucleotide synthesizer that we used previously to produce a radioactive probe for the insulin gene, but this time we will make two oligonucleotides: One of these will be identical in sequence to the 20–25 bases on the left end of our 376-base segment, and the other will be identical to 20–25 bases on the right end (the precise number of bases chosen for each primer is made on the basis of convenience and optimal function of the PCR procedure). These oligonucleotides have two other important design features. Each is identical to the sequence of one of the two strands, and the other is identical to the sequence of the complementary strand. The other important feature is that the 3' end of each oligonucleotide faces toward the middle of the 376 base DNA segment. Thus, if DNA polymerase used these oligonucleotides as "primers" for replication of the DNA strand opposite them, the new strand would be extended in the direction beginning at the margin of the segment and proceeding toward the middle. This is because DNA strands are antiparallel (p. 24) and because new bases are always added in the 5' → 3' direction with respect to synthesis of the new strand (see p. 31). The positions of the sequences that match the oligonucleotides and the orientation of the oligonucleotides are shown in Figure 14. The oligonucleotide synthesizer can produce incredible quantities of these 20–25 base fragments in highly purified

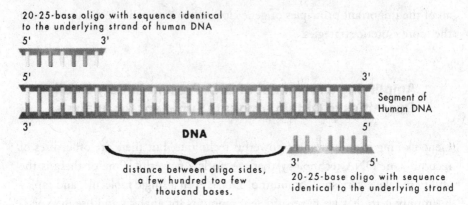

Figure 14   A small segment of DNA with the positions of short primers shown above and below each strand. The positions of the primers are several hundred to several thousand bases apart on the strand.

form. We will now mix the two oligonucleotides (which we will hereafter call "oligos" for short) in equal amounts and add them to human DNA such that for every copy of human genomic DNA there are several trillion copies of each of the oligos. This will give a preparation of double-stranded human DNA floating in a thick soup of single-stranded oligos.

What will happen now if we heat the mixture sufficiently to melt the human DNA and then cool it slowly? As emphasized previously (see p. 23), heating of double-stranded DNA followed by slow cooling results in very accurate restoration of the double-stranded molecule, with complementary strands returning to form hydrogen bonds with each other on the basis of the A:T/G:C principle. But in the mixture we've just created, something different will happen. The melting step will expose regions of human DNA that are exactly complementary to each of the oligos. This is true because we produced the oligos on the basis of our knowledge of the human DNA sequence and the opposite strand of the regions synthesized is of course perfectly complementary to each oligo.

Because there is a huge excess of the oligonucleotides relative to the human DNA present in the mixture, the chances are high that as the solution is cooled the oligos will intercede, thus displacing the human DNA and forming 22-base double-stranded segments on the extreme ends of the 376-base sequence around the mutation site in the globin gene that leads to sickle cell anemia. The appearance of the human DNA in our mixture after heating and cooling is shown in Figure 15. Once we have created the situation shown in this figure, we are ready to amplify the 376-base DNA segment by the PCR technique.

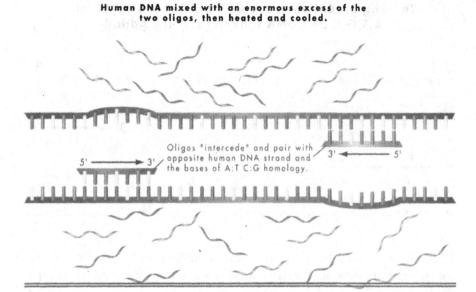

Figure 15 The same segment of DNA shown in Figure 14 after heating and slow cooling in the presence of a large excess of primers. Note that the primers displace the DNA strand by forming A:T G:C pairs with the complementary sequence on the opposite strand.

To perform PCR, we will add bases that can be used to replicate DNA and we will add DNA polymerase. When we do this, the DNA polymerase will recognize the free 3′ end of each oligonucleotide and, using the adjacent human DNA as a template, will begin the replication process, extending the oligonucleotide in the only direction it can, 5′ → 3′. If we allow this strand extension to continue for a minute or so, we have the result shown in Figure 16. Note that we have, over a short region between the oligonucleotides and perhaps a little beyond them on either side, replicated the human DNA. It is important to appreciate not only that we have replicated the region between the oligos, but that the replication has been allowed to continue such that the DNA sequences that correspond to the oligonucleotide primers (shown as P1 and P2 in Figure 16) are also replicated. This process has thus duplicated a very small region of the human genome, including the site where the oligos anneal, and we have used two oligos for incorporation into the new strand.

Let's now take this mixture and heat it again. The DNA will melt again, which will interrupt the DNA synthesis. If we cool this mixture yet again, however, we can now reanneal some of the remaining oligos to their complemen-

The Region Between the Oligo is Replicated When A,T,G,C and DNA Polymerase are Added

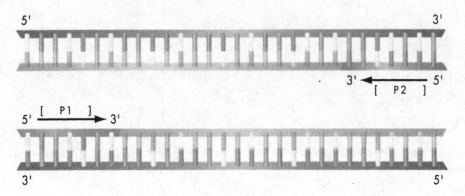

Figure 16 The same DNA segments shown in Figures 14 and 15 after completion of one cycle of PCR. Note that the region between *and including* the primer sites is replicated.

tary sites in the DNA. However, whereas we originally had only one site for oligonucleotide annealing, we now have two sites for each because the polymerase replicated the annealing sites along with the intervening DNA. The process of heating, cooling, and strand extension requires about 3 minutes and leads to an amplification of the DNA region flanked by the oligo sequences. This is one cycle of PCR.

PCR is exponential, such that one cycle produces two copies of the segment in question, two cycles of PCR produce four copies, etc. If we perform 35 cycles of PCR we still will not come close to using up our enormous excess of oligos, but we will have a tremendous amplification of our 376-base DNA region. So great has this amplification been that it is sufficient to visualize the DNA with the naked eye. Visualization is accomplished by loading the DNA mixture into a rectangular gel made of agarose, which is much like the agar used on our bacterial plates previously. The gel is then placed in an electric field, and the DNA moves into the gel toward the positive pole in the electric field. Once the billions of 376-base DNA fragments have moved into the gel, we can visualize them as follows. We place the gel in a solution containing a substance, ethidium bromide, that avidly forms a complex with double-stranded DNA. Ethidium bromide fluoresces in ultraviolet light, so after we expose the gel containing our amplified fragment to the ethidium bromide we can examine the gel under ultraviolet light and the fragment will fluoresce intensely.

So why did we take the trouble to amplify this DNA fragment? In the example given, we have amplified a region of the globin gene that includes the site of the sickle cell anemia point mutation. Suppose we had obtained the DNA sample by amniocentesis and the DNA is from a fetus that the parents are concerned may be born with sickle cell disease. We can very simply determine by PCR whether the fetus has the disease, is a carrier of the disease, or is normal. To show how simple this is, we will examine the precise bases that immediately span the mutation site. These bases are located 175 bases from the left end of our amplified fragment. The sequence of the key base altered in the disease and those immediately flanking it is CTGAG. When the sequence is mutated to give sickle globin, the sequence reads as follows: CTGTG. Recall that it is the GAG to GTG change that substitutes one amino acid into globin to produce sickle globin. An especially interesting feature of this sequence is that CTGAG is the recognition sequence for a restriction enzyme, *Dde*I, which cuts this sequence between the C and T as follows: C—TGAG. In addition, it should be readily appreciated that the sickle mutation alters this site such that *Dde*I will no longer recognize it. *Dde*I cuts CTGAG, *but not* CTG*T*G.

Because normal globin gene at the position of the the sickle mutation fortuitously contains a *Dde*I site, we can determine whether the mutation is present by simply digesting our PCR product with *Dde*I and determining whether the fragment is cut. In normal DNA *Dde*I will cut at this position to produce fragments of 175 and 201 bases from our 365-base PCR fragment, but in sickle DNA *Dde*I will not cut and the 376-base fragment will remain intact. To see whether the fragment cuts, we can load the *Dde*I-digested PCR product on a gel, place it in an electric field for an hour or so, stain the gel with ethidium bromide, and examine the gel under ultraviolet light. If the material is digested by *Dde*I there will be two fragments, 175 and 201 bases in size, whereas failure of *Dde*I to digest will yield a single fragment of 376 bases. When DNA fragments are run into gels with electric fields, the smaller fragments run faster because they can make their way through the agarose matrix more easily. So when our PCR product from unmutated DNA is digested with *Dde*I we will see two fluorescent bands, with the 175-base band further away from the loading site in the gel.

Remember that we said that our analysis could distinguish a fetus without the mutation from a fetus carrying the mutation and from a fetus that will develop the disease. The fetus without the mutation will have two bands, 175 and 201 bases in size after PCR, *Dde*I digestion, and gel analysis, whereas the fetus destined to develop sickle cell disease will have a single 376-base band. What about the fetus that inherits a sickle globin gene from one parent and an unmutated gene from the other and that is therefore a carrier of the sickle trait? The amplification product of the gene without the mutation will give the two bands expected, and the other gene will give the single band characteristic of sickle globin. Thus this sample will have three bands, the uncut band from the sickle gene and the two smaller bands from the gene without the mutation. Thus a carrier of the sickle trait can be easily and unequivocally identified. This analysis is shown in Figure 17.

The aforementioned example shows how PCR can be used to perform diagnosis of genetic disease in fetuses, but there are thousands of other uses of this technology. Some regions of DNA that encode genes involved in the immune response are virtually unique in every person. If a few cells are left behind at the scene of a murder, PCR can be used to amplify the minute quantities of DNA in these cells, with the oligonucleotide primers selected to amplify regions of the DNA that are highly variable between individuals. This kind of analysis

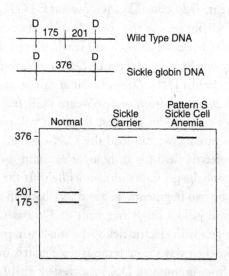

Figure 17 Analysis of sickle cell anemia by PCR and *Dde*I restriction enzyme digestion. In normal individuals all DNA is cut with *Dde*I. In carriers of the sickle mutation, DNA from the unmutated globin gene is digested but the sickle globin does not digest because of loss of the *Dde*I site. In individuals with sickle cell anemia or in embryos and fetuses destined to develop the disease, none of the DNA will digest with *Dde*I.

can provide unequivocal identification of the owner of those cells and can thus implicate or definitively exclude suspects in crimes. So powerful is PCR that even a single molecule of DNA can be amplified to produce products that are readily analyzed. The use of PCR to study the genes of embryos with only a few cells will be discussed again later in this book.

DNA Sequencing

The use of PCR for diagnosis of sickle cell anemia stands as one example of the importance of DNA sequencing to medicine. This PCR procedure could not be conducted without knowledge of the sequence of the bases in the immediate vicinity of the sickle mutation. One of the most important recent break-throughs in genetics has been the complete sequencing of the human genome. This achievement offers innumerable opportunities to diagnose genetic disease, to understand the contribution of genes to disease, and to understand the fundamental nature of the human species as well. In addition, this information can broaden the choices of individuals who might consider elective or therapeutic procedures such as cloning, which we would categorize as a form of germ line gene manipulation. For these reasons it is important to understand at least one method of sequencing. Fortunately, DNA sequencing is not hard to understand: It relies on our standby principle of A:T, G:C pairing.

To sequence a fragment of the human genome, we will first clone it into a plasmid as we did for the *Eco*RI fragment of the human insulin gene described above (see pp. 45–46). The sequence of the plasmid immediately flanking the cloning site can be readily determined and even created in the laboratory and added to the plasmid. To sequence the fragment inserted into the plasmid, we will produce an oligonucleotide that is able to pair with one strand of the plasmid in the region immediately adjoining the cloning site. This primer is quite similar to the PCR primers in that addition of bases and DNA polymerase will allow DNA synthesis to attach new bases to the primer. We will arrange this primer such that the synthesis of the new strand, which is always in the 5′ → 3′ direction, will extend from the flanking region of our cloned insert directly into the cloned fragment, and the DNA synthetic process will then use one of the strands of the cloned insert as a template for insertion of complementary bases. This arrangement is shown in Figure 18. In addition to producing a primer with the position and orientation that allow copying of one strand of our cloned insert, we will label the primer with radioactivity. Thus, if we pro-

DNA Synthesis for Sequencing

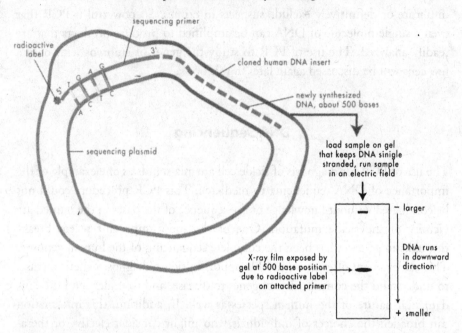

Figure 18   DNA synthesis for sequencing. DNA is cloned into a sequencing plasmid, and a radiolabeled primer is annealed just upstream from the human DNA insert for extension by DNA replication. After a period of strand extension, the product is run in a gel through an electric field and exposed to X-ray film. The position of the newly synthesized DNA, now linked to the radiolabeled primer, is seen by exposure of the X-ray film.

duce a 500-base strand by DNA synthesis, run this strand into a gel, and expose the gel to X-ray film, the film will be exposed at the position corresponding to where 500 base fragments run. This is shown in Figure 18. But how does synthesis of a new DNA strand with a radioactive tail tell us the sequence of the DNA that we have copied?

The way we can use DNA copying to determine sequence is by modifying some of our bases so that when they are inserted into our newly growing strand of DNA they block any further bases from being added. In effect, we may think of the "lock" portion of these bases as shown in Figure 3 to be permanently closed by chemical modification. We will set up four test tubes with the identical plasmid in all four, and we will copy our cloned insert by the identical method in all four reaction tubes. However, in each of the four tubes, a different base is modified such that its incorporation into the newly growing strand will block further bases from being added. In one tube the As will be "locked," whereas in each of the others the Gs, Cs, or Ts will be locked. The locked versions of these bases are known as "dideoxy" bases. It should be easy to appreciate that in our first tube, modification of the As means that the last base of any new strand synthesized will be an A. In our tube with the modified Gs, the last base added to any new strand will be a G, and so on. Let us now imagine that we run all four DNA synthesis reactions, load the products on a gel, run them into the gel, expose the gel to X-ray film, examine the film, and find that in tube 1 a 250-base band appears, in tube 2 a 251-base band appears, in tube 3 a 252-base band appears, and in tube 4 a 253-base band appears. Logically then, bases 250, 251, 252, and 253 are A, G, C, and T. This result is diagrammed in Figure 19. Actually of course, the bases on the strand copied were T, C, G, and A, because our new strand contains bases complementary to the copied strand. The only question now is, How were we able to obtain a 250-base strand in tube 1 when one of four bases in the tube (the As in this case) are modified such that DNA synthesis will be terminated? Of course, the strand being

DNA Synthesis Performed With Dideoxynucleotides for Sequencing

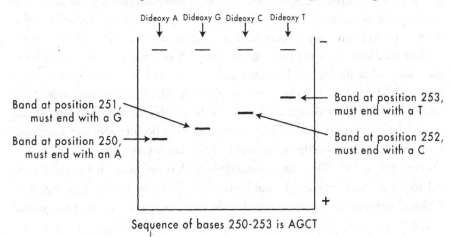

Sequence of bases 250-253 is AGCT

Figure 19   A sequencing gel visualized after performing 4 sequencing reactions, each with the dideoxy nucleotide derivative of a different base added to the reaction. Incorporation of the dideoxy base stops further strand extension, allowing the experimenter to determine the last base inserted into the strand. In this figure, bases 250–253 are shown in adjacent lanes.

copied might fortuitously not contain a T for 250 bases, but it certainly would have Cs, Gs, and A's. And yet, we were able to produce at least a 250-base strand in each of the tubes. What's going on? Why were the newly growing strands not stopped much sooner?

The way to get long strands synthesized when one of four bases is modified to terminate DNA synthesis is to modify only a fraction of the bases. So, in tube 1 we will modify 10% of the As and leave the other 90% unmodified. In this reaction tube new strand synthesis will always have A as its last incorporated base, because As are the bases modified to terminate strand extension. However, because not all As are modified we can sometimes put this base into the strand and continue strand extension. Actually, then, this reaction tube will produce a series of strands of different lengths, but every strand produced will have A as its last incorporated base. It should be clear that the greater the percentage of modified As we include in tube 1, the sooner one of the modified bases will be inserted and the shorter the new strand will be. By controlling the percentage of modified bases in each of the reaction tubes we can thus generate a "ladder" of bands on our X-ray film, and we can determine the order of several hundred bases by looking at the ladders in each of the four lanes that correspond to the four reaction tubes.

When this type of sequencing procedure is performed, one can readily obtain several hundred bases of new sequence information. To continue the process, one synthesizes a new primer based on the sequence deduced at the furthest point from the first primer. This new primer can then pick up the extension reaction where the previous reaction left off, and another several hundred bases of sequence can be obtained. A final point to appreciate regarding this methodology is that the newly synthesized DNA is single stranded, because our primer was designed to make hydrogen bonds with only one of the strands (this contrasts with PCR, where primers to both strands are made and arranged such that newly synthesized DNA crosses the space between the primers; see p. 53). When single-stranded DNA is produced, it has the potential to form short regions of double-stranded DNA by base pairing between different parts of the single-stranded molecule. For example, if one part of the strand had the sequence AATGGC and another was TTACCG, these two sequences could form hydrogen bonds with one another and cause the strand to loop back on itself, much as transfer RNA does by localized A:U, G:C pairing as shown in Figure 9. Such loops can affect the mobility of the DNA in the gel, so we avoid such problems by making the gel with solutions that make the formation of these hydrogen bonds impossible.

In recent years new sequencing methodologies and instrumentation have become available, and these advances have allowed far more rapid sequencing. These technological improvements have made a major contribution to the successful sequencing of the human genome. In addition to the use of sequence information to perform PCR and diagnose genetic disease, knowledge of the complete sequence of the human genome has many other important applications to medicine and biology that will be touched upon later when genetic engineering strategies are detailed.

Gene Splicing

Another application of recombinant DNA technology that is central to our discussion of engineering the human germ line is gene splicing. By using restriction enzymes, we can cut out fragments of one gene and substitute them for the corresponding fragments in another. For example, we can remove the promoter region from the insulin gene and replace it with the promoter region of the globin gene. It should be apparent from our discussion of plasmid cloning that all we need to do to accomplish this is to have conveniently situated restriction sites and a plasmid to work with. We can clone genes into plasmids, remove regions with restriction enzymes, and reclose the circle by adding fragments from other genes that have the same restriction enzyme ends as those produced when the plasmid was first cut. Now perhaps you may think such a project very challenging. How can we expect to be so lucky as to find restriction enzyme sites in just the places we need? In reality, such procedures are quite simple because there are very many restriction enzymes and because it is possible to modify ends produced by one restriction enzyme so they can anneal and be ligated to those produced by a different enzyme. Another source of flexibility in designing such strategies is that the addition or removal of a few bases in untranslated regions around genes often has very little effect on gene function. Thus, if the TATA box is 30 bases away from the transcription initiation site of one gene and it is grafted to a position 35 bases from the transcription initiation site of another, it often will still work perfectly well.

These examples of the use of recombinant DNA technology have special relevance to the field of germ line genetic engineering. When we discuss the various strategies for modifying the germ line, these techniques will come into play.

Gene Splicing

4

TRANSMITTING THE GENETIC INFORMATION TO FUTURE GENERATIONS

The course of development consists simply in this; that in each
generation the two parental traits appear, separated and unchanged,
and there is nothing to indicate that one of them has either
inherited or taken over anything from the other
—Gregor Mendel, in a letter, April 18, 1867

As discussed in Chapter 1, the ultimate act of rebellion against the law of entropy is reproduction. Because it is physically impossible for a living entity to maintain its state of organization forever, it does the next best thing: It reproduces, to create other living beings that are much like itself. When single cells are involved, the reproduction process simply involves the production of new cells by cell division. When a multicellular organism like a human is involved simple division is not possible, so other more elaborate approaches must be used. Instead of dividing, we have children. In this chapter we will see how replication and reproduction take place. The essential step in both approaches is ensuring that the new cells or organisms produced receive the correct genetic information in the correct amount. Without an intact body of genetic information the offspring, of course, cannot function properly. We will first see how cells ensure transmission of the complete lexicon of genetic information in the

The Science and Ethics of Engineering the Human Germ Line: Mendel's Maze, by Jon W. Gordon
ISBN 0-471-20647-4 Copyright © 2003 John Wiley & Sons, Inc.

process of cell division, and we will then examine how big colonies of cells like humans accomplish this objective.

Cell Division—Asexual Reproduction

Cells that exist as free-living organisms like protozoa reproduce by simply dividing. Precisely the same process is used for cells that multiply in a multicellular organism. This method of gene transmission does not involve joining of genetic material from two genetically distinct cells, and it is therefore referred to as asexual reproduction. The division process is called mitosis, and here's how it works.

We previously made an analogy between the nucleus of a mammalian cell and a computer hard drive. In addition to containing an amount of information similar to that on a hard drive, the nucleus of a cell uses a similar strategy for storing that information so that it can be efficiently accessed. As you know, the data on your computer hard drive is distributed into sectors—blocks of data that carry the total code. The nucleus also stores its genetic information in sectors, which are called chromosomes. The DNA that encodes a human being is divided into 23 chromosomes, and as noted above, each cell has two complete copies of the information. Thus a human cell has two sets of 23 chromosomes, to give a total of 46. Because 23 chromosomes carry 3 billion bytes of information, represented biologically as DNA bases it follows that 1 chromosome may have several hundred million bases of DNA, and this is in fact the case. A chromosome can be thought of as a single long strand of DNA. In chromosomes the DNA is complexed with protein and RNA molecules (not mRNA), to protect it and help keep it bunched tightly so its meter of length will fit in the microscopic nucleus. Normally chromosomes are not visible, but they can be seen under the microscope when the cell divides. We will examine the distribution of DNA to offspring not as 23 chromosomes, but rather, for the sake of convenience, only 3 chromosomes. When we consider sexual reproduction—the human method of "replication," we will again consider a hypothetical case in which the organism has only three chromosomes in each of its cells.

Figure 20A shows the chromosomes within the nucleus of our model cell. In this example I have simplified the situation to depict a "humanoid" male cell with an X chromosome, a Y chromosome, and two nonsex chromosomes, or "autosomes." Note that the red autosome is large and the green one is much smaller, just as some hard drive sectors may contain more data than others. To

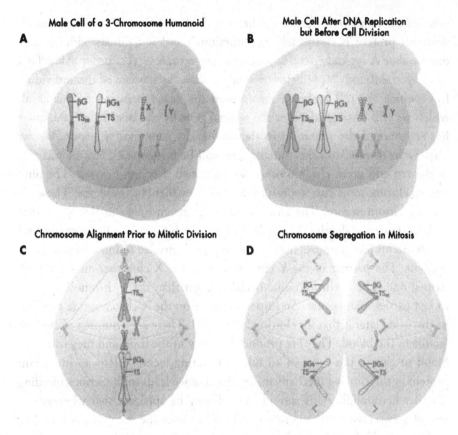

Figure 20   Cell division (mitosis). The DNA complement shown as the chromosomes is shown in A. DNA is first replicated (B); then the chromosomes align at the equator of the cell (C) and split down the middle (D). The result is two daughter cells with the exact chromosome composition of the parent cell.

distinguish the two sets of chromosomes, one of the chromosomes from each set is solid while the other is striped. The solid chromosomes represent the data set passed to this model organism from the father, and the striped chromosomes are inherited from the mother. Both sets contain all of the genetic information, so the organism has two complete copies of the data set. The red striped chromosome accordingly contains the same genes as the solid, but is striped so that we can follow each member of the set as we proceed through replication. To assist us further in our analysis of cell replication and sexual reproduction, we will track the fates of two genes on the large chromosome. One is for the β-globin gene, which we will abbreviate βG, and the other is the gene that is mutated in Tay–Sachs disease, which we will call TS. The Tay–Sachs

gene encodes an enzyme known as hexosaminidase A. In the Tay–Sachs mutation this protein is completely nonfunctional, and individuals without hexosaminidase A typically develop a devastating neurological disorder that is fatal
in early childhood. Note before we begin that the mother of the individual
from which the cell is shown transmitted the sickle mutation to this child,
which we will distinguish from the normal gene by the label βGs. The father
has transmitted the mutation in the Tay–Sachs gene that is responsible for
Tay–Sachs disease, and this mutant gene will be designated TSm. In addition
to the red and green chromosomes, we have two that don't match. The pink
striped chromosome corresponds to the much smaller blue solid one. These are
the sex chromosomes. The small solid blue chromosome carries genes that
make a conceptus develop as a male. In humans this is the Y chromosome, and
for all intents and purposes it contains only genes needed to induce male development. The partner of the Y chromosome is the X chromosome—the pink
striped one. The X chromosome is much larger than the Y chromosome, and
in fact carries many genes. An important gene on the X chromosome that will
be discussed later is that for clotting factor VIII. There are a number of clotting
factors in the blood. They are produced mainly in the liver, and they cause the
blood to clot in the event of an injury. Clotting factor VIII is an important
protein in the clotting mechanism, and its absence leads to the serious bleeding
disorder hemophilia A. It should immediately be apparent that whereas two
sets of genes exist on the chromosomes other than the sex chromosome, the
genes on the X chromosome have no counterparts on the Y chromosome. Thus
there is only one copy of the genes on the X chromosome in this "male" cell.
Before we proceed to replicate the cell we should be aware that a cell from a female of the same three-chromosome species contains two X chromosomes and
no Y chromosome. In females, therefore, each cell does in fact carry two copies
of genes on the X chromosome.

An important point to appreciate is that although the same genes are present
on the solid chromosomes and their striped counterparts, the genes might not
be exactly identical. In our example, the globin gene present on the large solid
chromosome, inherited from the father, is the more common "normal" gene,
whereas the globin gene on the striped chromosome carries the base change
that produces sickle globin (see pp. 36–37). In this case the DNA base sequences of the globin genes at their corresponding positions on each of the
chromosomes are almost but not exactly identical: Although both globin genes
produce protein, the amino acid sequences of the globin proteins produced
from the two genes differ by a single amino acid—at the sixth position of the

amino acid chain (see pp. 37 and 56). We therefore would have, in this cell, two different versions of the globin gene. Similarly, the TS genes from the father (solid chromosome), where the Tay–Sachs disease mutation resides, and that of the mother, which is the normal TS gene, differ slightly. Different versions of the same gene are referred to as "alleles" of the gene.

Another feature of the chromosomes that we have not yet mentioned is the solid, circular swelling present near the middle of each chromosomes. This region, called the centromere, plays a very important role in ensuring correct transmission of the genetic information to daughter cells during the process of cell division. The centromere is made of DNA, because it is part of the long strand that comprises the chromosome. However, the DNA sequences in this region are specialized such that they cause the centromere to form in this region.

The process of cell division results in one cell giving rise to two new ones, and the objective is to have the new ones be as much like their parent cell as possible. To accomplish this purpose it is critically important that each of the two cells produced from the parent cell have exactly the same genetic information as the parent cell. In our "mock" cell, this means that each new cell must get a solid and a striped red chromosome and each must get a solid and a striped green chromosome. Of course, each cell must also get an X and Y chromosome. If two new cells are each going to receive a full set of chromosomes, and there exists only one set in the parent cell, it will obviously be necessary to copy the genetic material before cell division. We have already described the DNA replication mechanism (see pp. 30–31) and pointed out how rapidly new DNA is synthesized in dividing cells. Let's take a look at our cell after DNA replication, but before cell division. This situation is depicted in Figure 20B.

Note now that each chromosome has two "arms" each instead of one and that these arms remain attached to one another at the centromere. Where there were previously two copies of βG and TS, we now have four (only two are labeled). What must be done now is to distribute a copy of the genetic material to each daughter cell that is identical to that found in the parent cell. To accomplish this task the centromere attaches to an elaborate protein complex in the cell termed the mitotic spindle. Basically, the spindle anchors itself at both sides of the cell and sends wirelike proteins toward the equator of the cell, where they attach to the centromere. Figure 20C shows the cell with its chromosomes replicated and lined up at the equator and connected to the spindle apparatus. The anchors for the spindle at each side of the cell are shown as little cylinders of protein. These cylinders are called centrioles. It should also be clear that if chromosomes are going to be separated and sent to two daughter cells, the nuclear membrane must break down. Nuclear membrane breakdown occurs after the DNA replicates, as the chromosomes prepare to line up at the equator and attach to the spindle.

When the cell begins to actually split, the chromosomes separate at the centromeres and are "dragged" away from the equator. A new membrane forms between the sets of chromosomes, and two daughter cells are produced that have genetic complement identical to that of the parent cell. This separation step with the early formation of the new cell membrane is shown in Figure 20D. As the new membrane forms, a nuclear membrane also forms in each new cell to provide a compartment for its genetic information. When the process is complete, the two daughter cells are genetically identical to the parent cell. Cell division, or mitosis, is accordingly quite straightforward. This is the method by which free-living single-celled organisms reproduce, and it is the method by which the fertilized egg multiplies to produce the many cells of the adult. Now we will turn our attention to sexual reproduction by the adult, a slightly more involved process.

Sexual Reproduction

When sexual reproduction takes place, the priority is the same—the new organism must be constructed so as to have a complement of genes identical to that of the parent organism. However, now there are two parents. If a father and mother both donated the full complement of genes found in each of their adult cells to the new offspring, the cells of the newborn would have four copies of every gene because the cells of the parents each have two full copies of the genetic code. If the newborn had twice as many genes as its parents, there would be big trouble. After all, what would happen when the offspring had their own progeny in turn? Would the next generation then have eight copies of the genetic code per cell? When would this genetic expansion ever cease?

Clearly, sexual reproduction—the creation of new offspring that inherit genetic information from two different parents—will only succeed if each parent donates only some of its genetic information, so that when cells from each parent combine there will be exactly two copies of every gene, just as in the parents. Logically, each parent will donate half of its genetic information, and the simplest approach to this challenge is to select one of each of its two genes in the genome and include it in the reproductive cell. The key issues in sexual reproduction are the following: First, if each parent produces reproductive cells that carry only half of the genetic information, how does the organism make sure that the half imparted to each reproductive cell actually contains one copy of every gene? Second, if each parent is going to select one of its copies of each gene for inclusion in a reproductive cell, which one of the two will it select? This last question is quite important when we remind ourselves of the concept of allelism: In the case where one globin gene is normal and the other carries the sickle mutation, will the reproductive cell get the normal gene or the sickle gene? Third, and finally, we might ask why the organism is going to all of this trouble to reproduce sexually when it is so simple to divide by mitosis as the free-living single-celled organism does?

Let's deal with the last question first. When organisms exist as huge multicellular colonies, it's difficult to envision a method whereby the colony would simply duplicate itself. Second, sexual reproduction offers some significant survival advantages for the species. If we think again about the globin alleles, one copy being normal and the other carrying the sickle mutation, and we realize that the reproductive cells, or gametes, cannot carry both alleles, it becomes apparent that the offspring will not be likely to carry the same pattern of alleles as either parent. For example, if one parent, say the mother, carries two sickle globin genes, and the father carries two normal genes, the offspring will receive a gamete with a sickle gene from the mother and a normal gene from the father. This reality makes it inescapable that, with respect to its two globin genes, the offspring cannot be genetically identical to either of its parents: The offspring will have one sickle and one normal globin allele, whereas one of the parents has two sickle globin alleles and the other two normal alleles. When one extends this principle to the 40,000 genes that the offspring inherits from each parent, it should be easy to predict that the child has no chance of being genetically identical to either parent. Although it will have the same genes in the same numbers, its pattern of alleles for those genes will differ. Of course, we are really stating somewhat formally what most of us already know—children are similar to but not identical to their parents. As we shall see when we look at sexual reproduction more closely, the selection of one of each chromosome pair to impart to each gamete is only one mechanism by which alleles are shuffled.

The generation of genetic diversity brought about by sexual reproduction contributes importantly to the survival of the species. Although it of course falls a little bit short of replication as a method for maintaining the state of organization that characterizes the parent, it's still pretty good and offers advantages to the species as a whole. Genetic diversity created by sexual reproduction makes it more likely that a combination of alleles present in a child will allow that child to survive the rigors of the environment. Consider the case of the carrier of sickle cell anemia, like the three-chromosome child in our hypothetical example. On the face of it, such a mutation would appear harmful, and given the choice, you would probably elect not to carry such an allele. However, as it turns out, carriers of a sickle cell allele are resistant to malaria, which is caused by a parasite that likes to multiply in blood cells. Blood cells produced with some sickle globin present a hostile environment for the malaria parasite. Thus, from the point of view of survival of the human species, it's good that in regions of the world where malaria is rampant some people carry the sickle allele. The price paid for this advantage is that some individuals may carry two sickle alleles, and these people will be affected with a serious disease, sickle cell anemia. The purpose, then, for sexual reproduction, is to come as close to direct replication of the organism as possible and to compensate for imperfect replication by generating genetic diversity, which helps preserve the species as a whole. Let us now examine sexual reproduction at the level of the chromosome.

The generation of sperm and eggs, which will contain half of the amount of genetic material as the cells from which they form, begins in the same way as does mitosis: with DNA replication. Although it might seem odd that the first step in reducing the amount of genetic material in the cell begins with a doubling of that material, the process compensates for this "misdirection" by replicating once and dividing twice. Here's how it's done.

Figure 21 shows a cell destined to produce sperm (remember we're dealing with an XY cell) after replication of the DNA and alignment of the chromosomes at the equator (upper left panel). The spindle has attached to the centromere in preparation for movement of the chromosomes to opposite sides of the dividing cell. However, a closer look at the chromosome alignment and attachment of the spindle apparatus reveals critical differences from what occurs in mitosis. In mitosis, all duplicated chromosomes are lined up vertically along the equator and attached to spindle fibers emanating from both poles of the cell (see Figure 20C). In the first step of meiosis—the process whereby the amount of DNA is ultimately halved—the solid red chromosome is aligned "side by side" with its counterpart, the striped red chromosome, and is attached to the spindle fibers emanating from the left pole only, whereas the striped red chromosome is attached to fibers from the right side. In the case of the smaller green chromosome, it is the striped chromosome that is attached to the left spindle and the solid chromosome that is attached to the right side.

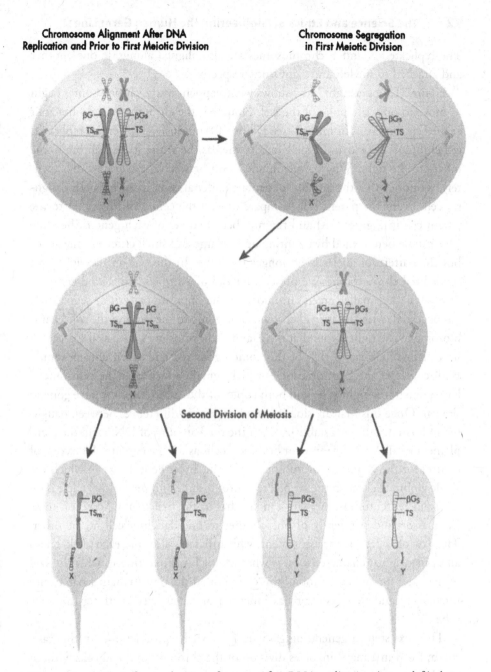

Figure 21 Meiosis for production of sperm. After DNA replication (upper left) chromosome pairs align at the cell's equator. When one pair goes to one daughter cell, the other homologous pair always goes to the other cell in a process termed "independent assortment" (center). The second meiotic division proceeds as in mitosis, but without DNA replication. The result is four cells with half of the DNA complement of the parent cell before DNA replication (bottom).

The replicated X and Y chromosomes are also aligned alongside one another and attached to the left and right sides, respectively.

Figure 21, upper right, now shows what happens as the chromosomes begin to move to opposite poles of the cell. Note that the chromosomes do not split at the centromere as in mitosis (see Figure 20), but instead the entire replicated red chromosome moves to the left side whereas the entire striped chromosome moves to the right. When a membrane forms between the two new cells (center), we notice that the amount of genetic material is the same for both daughter cells as for the parent cell (compare cells in the center of Figure 21 to the parent cell in Figure 20A) and the number of copies of each gene is the same (this can be determined by counting up the large and small chromosome arms) but the pattern of alleles is no longer the same. In the cell on the right, two striped red chromosomes are present, instead of a striped and a solid chromosome. For the smaller green chromosome the opposite pattern of segregation has occurred, with the striped chromosome going to the left and the solid chromosome to the right. Note that the X and Y chromosomes have also segregated in a manner similar to that of the homologous red and green chromosomes: Both copies of the Y chromosome, which remain attached at the centromere, have gone to the right side, and both copies of the X chromosome have gone to the left. Once this first division is complete genetic differences between daughter and parent cells are established, but the total amount of DNA and the complement of alleles is the same for both new cells as for the parent cell (except, of course, for the X and Y chromosomes). In this hypothetical case in which the sickle globin allele is on the striped red chromosome and the normal allele is on the solid red chromosome, you can see that in the first meiotic division these traits assort independently of one another to one daughter cell or the other. This law of independent assortment, which dictates also that each allele retain all of its original characteristics, is what Mendel was referring to when he said in the letter quoted at the top of this chapter that the traits remain separate and unchanged and that "neither has inherited or taken over anything from the other."

The next step in generating gametes for sexual reproduction—in this case, sperm from in a male, involves division of the chromosomes and cells without another round of DNA replication. For this second division of meiosis, the chromosomes line up vertically along the equator of the cell just as they do in mitosis, with spindle fibers from both sides attaching to each chromosome. Remember that these chromosomes still represent replicated DNA with two centromeres because the replicated chromosomes moved intact to one side or the other during the first division. In the second division of meiosis they split as in mitosis, and one copy goes to each side as two cells form from each of the two parent cells. We now have produced four cells from the original parent, and the amount of DNA is half of what it was in the original parent before meiosis. To appreciate this, it is necessary only to count the number of double-stranded DNA molecules in the original cell and its four offspring, recalling that each chromosome is a double-stranded molecule of DNA. In biology jargon, each double-stranded DNA molecule is called a chromatid, so that each chromosome has one chromatid before DNA replication and two chromatids after replication but before splitting. In the four cells resulting from meiosis we have one copy of every allele not on the X or Y chromosomes, but, as you can see, two of the four cells carry the βGs allele of globin and the other two carry the βG allele. Similarly, two cells have the TSm mutant allele and two do not. It should also be clear that two of the four cells, those carrying the Y chromosome, will produce a male conceptus at fertilization and the other two will lead to a female conceptus. To emphasize that each of these products of meiosis will become a sperm, we have added a tail to each one.

In the example we have given for the first division of meiosis, the striped red chromosome segregated with the solid green chromosome and with the X chromosome. However, in nature it is just as likely for the solid chromosomes to have segregated together to one cell in the first meiotic division, with all striped ones going to the other cell. The decision as to which chromosomes will go with which is random. If we consider for a moment the two chromosomes that are not the sex chromosomes (we call these the autosomes), we can appreciate that four different patterns of segregation could occur: small striped with big striped, small striped with big solid, small solid with big solid, and small solid with big striped. When there are two chromosomes as in this case, the number of possible segregation products is 4, which is 2^2. Without getting bogged down with the math, suffice it to say that for any number of chromosomes Z, the number of possible segregation products is 2^Z. Consider what this means for the generation of genetic diversity in meiosis. In humans there are 23 chromosomes and therefore 2^{23} possible segregation products in the first meiotic division. This is 8,388,608 different patterns of chromosome segregation! If we appreciate that many genes have allelic variants much like globin, we realize that each of these combinations gives a genetically unique pattern of alleles. Of course, this independent assortment of chromosomes is going on in both sexes, so 8,388,608 different possible eggs or sperm can be produced in each parent. Because any one of these sperm is equally likely to join with any of the eggs at the moment of fertilization, the number of possible complements of alleles, when considering just the phenomenon of independent assortment of chromosomes in meiosis, is 7,068,744,177,660—about a thousand times more than the total world population. It is therefore not surprising that siblings are never genetically identical unless they are derived from a single fertilized egg.

When eggs are produced, meiosis is the same in principle but slightly different in process. First, of course, the parent cell has two X chromosomes, so all mature gametes will carry the X chromosome and none will have a Y chromosome. Second, when the cells divide in meiosis in the female, the cytoplasm splits very unevenly: One set of chromosomes goes to a very large cytoplasmic compartment and the other to a tiny "polar body" that, because it is so small, will not go on to produce a functional gamete, or egg. In the second division of meiosis in the female the same uneven splitting of the cytoplasm occurs, such that a large functional cell and a small nonfunctional polar body are produced. Because of this uneven division of cytoplasm and "loss" of sets of chromosomes into tiny polar bodies, completion of meiosis in the female results in only one functional gamete instead of four. This is illustrated in Figure 22. Another unusual feature of meiosis in the female is that it is not completed until after the sperm enters the egg. The first division of meiosis is completed as the egg matures, and at the time of fertilization, the chromosomes are lined up in the egg in preparation for the second division of meiosis. The cell has no nucleus because the membrane has broken down in preparation for division of the chromosome into two separate cytoplasmic compartments. Completion of this second division and generation of the second polar body only occurs after sperm entry.

As mentioned above, the independent assortment of chromosomes in the first meiotic division generates more potential genetic diversity among offspring than any couple could ever need. But, believe it or not, there is another mechanism at work that increases genetic diversity even further. To understand this mechanism, it is useful to return to Figure 21, upper left panel, where the chromosomes are aligned before separation in the first division of meiosis. We will focus on the large red chromosome.

Recalling that each chromatid is a double-stranded molecule of DNA, we see that these two aligned chromosomes have a total of four chromatids because, after the initial DNA replication, there are two strands of DNA in each of the two chromosomes. If we take a close-up view of these aligned chromosomes, we can easily appreciate that two of the chromatids lie adjacent to one another and two others are at the periphery of this paired complex. Note that

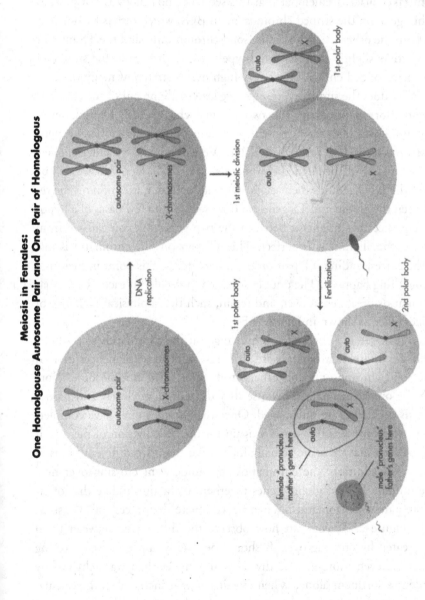

**Meiosis in Females:
One Homolgouse Autosome Pair and One Pair of Homologous**

autosome pair

X-chromosomes

DNA replication

autosome pair

X-chromosomes

1st meiotic division

auto

x

1st polar body

auto

x

Fertilization

auto

x

1st polar body

female "pronucleus" mother's genes here

auto

x

male "pronucleus" father's genes here

auto

x

2nd polar body

Figure 22 Meiosis for production of eggs. The processes of DNA replication and chromosome segregation are the same for eggs and sperm (compare with Figure 21), but in eggs cells divide unevenly, producing a first "polar body" (lower right) and, after fertilization, a second polar body (left). Meiosis in eggs is thus not completed until the father's genetic material has entered.

75

the chromosomes are aligned on the basis of sequence homology between them. The globin gene on the striped chromosome sits right next to the globin gene on the solid chromosome, and the same is true for the TS genes. In fact, it is on the basis of DNA sequence homology between the chromosomes that the alignment takes place. Remember that this sequence homology is not perfect: The globin gene on the striped chromosome is βGs, which differs by one pair of bases from the other βG gene on the solid chromosome, and the TS mutant gene also differs slightly from its "wild-type" counterpart. Still, the similarity between these aligned chromatids is very high over a stretch of many millions of bases. Why does the similarity of sequence lead to alignment of the chromosomes, and more particularly, the two internal chromatids? The answer is presently unknown, but this alignment phenomenon prevails in many experimental settings as well, as we shall see later. We will now take a much higher-magnification view of the aligned chromatids; so high, in fact, that we will be able to see the DNA strands themselves. We will focus our attention on the region between the globin and TS genes, and we will show a few bases of a hypothetical sequence that might sit between the two genes. As you can see from Figure 23A, the alignment is perfect. The AT pair on one chromatid is just next to the corresponding AT pair on the other, etc. At this point in meiosis a remarkable thing happens. The double-stranded molecules in each of the adjacent chromatids break, cross over, and rejoin, such that a physical exchange of DNA takes place as shown in Figure 23B.

We can now see the effect of this exchange on the chromatids as a whole. Note that the internal chromatids are "hybrids" with some sequences derived from the solid chromosome and others from the striped chromosome. Note also that the outer chromatids, because they did not align closely and participate in this exchange, are unchanged. Once the chromosomes have completed this crossover process and separate again for the first division of meiosis, it can now be seen that there are actually four rather than two different kinds of chromatids. In addition, the chromatids that underwent crossing over now have the mutant TS and globin genes together, a situation unlike that of either of the parents' chromatids. When we complete the process of meiosis as shown in Figure 23C, you can now observe the differences between them that are created by crossing over. It should be readily apparent that crossing over generates even more genetic diversity in gametes than was achieved by independent assortment alone. When crossing over is included in the calculations, two parents are able to produce several orders of magnitude more genetically unique children than the seven trillion or so made possible by inde-

Chromosome Alignment After DNA
Replication but Prior to Crossing-Over

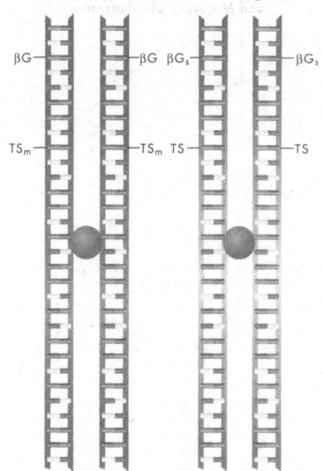

Figure 23 Chromosome crossing over in meiosis. The process is marked by the globin and Tay-Sachs genes. A. Chromosomes aligned after DNA replication but before crossing over. (*continued*)

pendent assortment and union of the gametes. This point is illustrated in Figure 24 by showing the result of crossing over in a single autosome before generation of sperm. Note that the crossover even increases the genetic diversity of the products of meiosis.

The enormous increase in genetic diversity created by crossing over is due to the rather substantial degree of variability in DNA sequences. In the model or-

Crossing Over Between Adjacent Chromatids on Homologus Paternal and Maternal Chromosomes

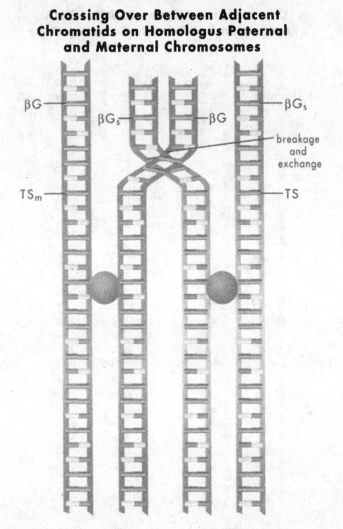

Figure 23   (*cont.*) B. A crossover event between the globin and Tay-Sachs genes.

ganism we have been using to illustrate the process of gene segregation in meio sis, the globin genes differ by one base, as do the normal and mutant TS genes Slight differences in sequence between the fathers' and mothers' genes are actu ally more the rule than the exception. This point can be easily appreciated by simply looking at one's own parents. One of them almost certainly has slightly different hair, eye, and skin color, a slightly differently shaped nose, etc. Be cause each of these traits is encoded by corresponding genes in each parent they must all differ slightly. This reality ensures that the exchange of pieces o

Product of Crossing Over, Completed Prior to First Meiotic Division

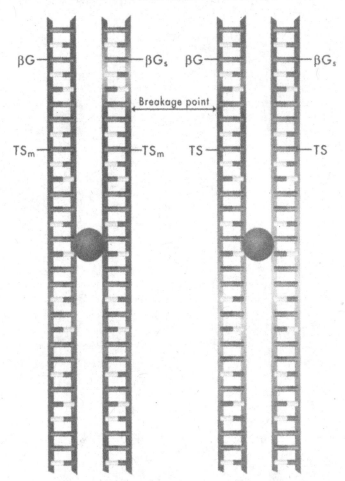

Figure 23 (*cont.*) C. Crossing over complete. Note the physical exchange of DNA between the participating chromatids and the shuffling of the pattern of alleles for globin and Tay-Sachs on these chromatids.

chromosome as shown in Figure 24 results in a significant reshuffling of association, or "linkage" of alleles. Moreover, crossing over is a very frequent event, with the chance of its occurring equal to about 1 percent for every million bases that separate genes. Because a chromosome can have more than 100 million bases of DNA, multiple crossover events can occur in such a chromosome during every meiotic division. Chromosome crossing over just before the first

Crossing-Over Increases Diversity in Products of Meiosis:

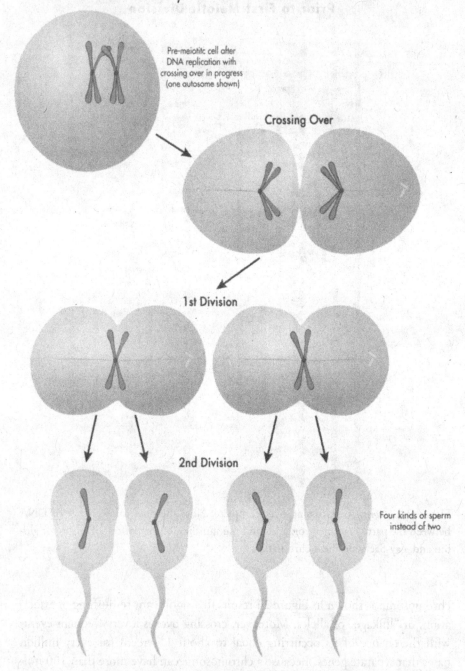

Pre-meiotitc cell after
DNA replication with
crossing over in progress
(one autosome shown)

Crossing Over

1st Division

2nd Division

Four kinds of sperm
instead of two

Figure 24 Crossing over during spermatogenesis in an organism with a single autosome, showing how crossing over increases genetic diversity.

division of meiosis therefore dramatically increases the genetic diversity over what is provided by independent assortment. The result of all of this process is that all offspring have the identical number of genes as do their parents, and they have two genes for every protein (except, of course, for genes on the X chromosome in males, see Figures 20 and 21). However, no two offspring have the exact pattern of alleles as either parent or each other. Another important point is that the regions of DNA between genes also vary slightly on the different parental chromosomes. In fact, these regions that do not encode proteins are more variable in sequence than are the genes themselves, where many changes can lead to death of the conceptus. You may wonder what difference it makes if sequences vary in regions where there are no genes. The biological significance of these DNA regions are not completely understood, but as we shall see later, variability within them can allow us to "track" regions of chromosomes as they are passed from parent to offspring. Just as it is possible to determine that a sickle globin gene came from one or the other parent, so it is possible to use PCR to identify noncoding sequences that are specific for a chromosome from one or the other parents. The ability to assign these "anonymous" sequences to a specific parental chromosome has important implications for genetic engineering. A final point to keep in mind is that although variations in sequence between homologous regions of the mother's and father's chromosomes is substantial, long regions of sequence identity between the chromosomes still exist, which enable the chromosomes to align for crossing over.

The Relationship Between Genotype and Phenotype

The "genotype" of an individual is defined as the specific pattern of alleles that an individual inherits. The genotype of a child who inherits βGs from one parent and βG from the other is a carrier of the sickle trait. With respect to the globin gene, such an individual can produce two different types of gametes, and the embryos, or "zygotes" that are produced after fertilization by these gametes can differ at this genetic locus. Because the zygote types that can be generated from such a parent are heterogeneous, the parent is called a "heterozygote" with respect to his/her globin genes. An individual with two identical alleles at a genetic locus is called a homozygote.

With respect to the heterozygote for the sickle gene it is reasonable to ask whether the individual will have sickle cell anemia, have no blood abnormality whatever, or have a mild disorder consistent with the fact that half of the globin protein is normal and half has the sickle mutation.

The answer to this question depends on the particular gene. In the case of globin, it is typical for carriers of the sickle mutation to have mild abnormalities that can be traced to the presence of one βGs allele in each blood cell. However, in many cases, mutation of one allele, even if it totally destroys the protein (as for example, might occur in a frame shift mutation as shown on p. 36), goes completely unnoticed as long as the other allele is normal. An example of the situation is Tay–Sachs disease. If an individual inherits a mutated Tay–Sachs gene from both parents and has no normal copies of the gene, the fatal disease results. However, inheritance of one normal and one mutated allele at the Tay Sachs genetic locus leads to no abnormality whatever. This second scenario is quite common. The third possibility is that inheritance of a mutated allele at a particular genetic locus will lead to full-blown disease, even though a normal allele has been inherited from the other parent. A good example of this pattern of inheritance is Huntington disease, a disorder leading to fatal degeneration of important regions of the brain in middle age. This latter pattern of disease inheritance is known as dominantly inherited disease, and the Huntington mutation is called a dominant mutation. The type of inheritance seen in Tay–Sachs disease is known as recessive inheritance, because the mutation does not manifest itself at all when a normal allele is present. Cases like the sickle globin gene are semidominant, with a partial effect of the mutation discernible. However, in these cases the abnormalities are so slight that such mutations are often considered recessive. The assumption implicit in this discussion, of course, is that both alleles at each locus are expressed equally. It is important to recognize that recessive mutations can cause disease only if the mutation is inherited from both parents and is present on both chromosomes of the child. When a parent is a carrier of such a mutation, half of his/her gametes will carry the mutant allele, as is apparent from examining the generation of sperm in meiosis (see Figure 21). When both parents are carriers of a mutation, the chances are 50:50 that a sperm with the mutation will reach the egg and are 50:50 that the egg that is fertilized will carry the mutation. Because half of a half is one-fourth, recessive disease is inherited from two carrier parents by 25% of the offspring. In dominant disease, only one copy of the mutated allele need be present for disease to appear. So, if one parent carries the gene (and, of course, has the disease), half of his/her gametes will carry the mutant allele and half of the offspring will have the mutant gene and develop the disease.

We cannot leave the subject of genetic inheritance without returning to the unusual situation of the sex chromosomes. Consider the male, who has a gene on the Y chromosome that determines male development of the fetus. On the X chromosome, however, there are hundreds of genes, some of them house-keeping genes that are indispensable for life. In males there is only one copy of these important X-linked genes, and in females with two X chromosomes there are two copies of these genes. This disparity raises two important questions. First, if every gene is expressed at an appropriate level for the organism to de-velop and function normally, how does the developing embryo deal with the fact that females have the potential to express twice as much protein from X-linked genes as males? Second, what happens when a serious mutation affects one of the genes on the X chromosome and this mutated gene is passed to a male, who has no other X chromosome in the cell that would compensate for the mutation by providing a second normal allele?

First, it should be understood that more gene expression is not necessarily better. Over 30 million centuries or so of evolution, levels of gene expression for an organism as complex as a human become very finely tuned. As a conse-quence, too much expression of genes can have disastrous consequences. For example, the disorder Down syndrome, characterized by mental retardation, heart defects, and a variety of other abnormalities that greatly shorten life ex-pectancy, is most often due to an error in chromosome segregation at the first meiotic division that results in the presence of three chromosomes 21 instead of the normal two (remember that the normal human cell has 23 chromo-somes). These disorders are associated with the 50% increase in expression of genes on this chromosome, a result of the fact that there are three copies of every gene instead of two. So females should not necessarily rejoice at the prospect of expressing twice as much of their X-linked genes as males. One might imagine that, because two copies of each gene is normal for genes not on the X chromosome, females, with two X chromosomes, are normal and males must compensate for their deficiency by doubling the activity of their X-linked genes. "Doubling activity," at the molecular level, means producing mRNAs at twice the normal rate or stabilizing the mRNA such that each one produces twice as much protein.

Although these strategies for compensating for the difference in "dosage" of X-linked genes between the sexes makes perfect sense, that is not how the ad-justment is made. Instead, the female cell cuts its activity of X-linked genes in half. How is this done? The embryo takes no chances. Early in development, one of the X chromosomes in each cell is chosen at random and condensed into a very tight ball that does not allow access to any of its genes. This condensed chromosome sits up against the inside wall of the nuclear membrane. Whether the X chromosome chosen for inactivation is inherited from the father (brought in by the sperm) or the mother, this same X chromosome is always the inactive one in subsequent generations of cells. That is, this inactive X chromosome is replicated and segregated in mitosis as the cell divides but is inactivated in both daughter cells once division is complete. As a result, female cells are much like the XY cells of the male—they have active genes from only one X chromosome. This process, which equalizes functional gene dosage for the X chromosome between the sexes, is called dosage compensation.

If human cells, be they XX (female) or XY (male), have only one set of active X-linked genes, what are the consequences of mutation of these genes? To consider this question we will return to an important X-linked gene, that encoding clotting factor VIII, and imagine that it has sustained a mutation that results in the absence of production of normal factor VIII. The consequences of such an event are actually quite different in males and females. Let's first take a look at males.

If a female with the factor VIII mutation produces an egg with this X chromosome and the egg is fertilized by a sperm carrying a Y chromosome rather than an X chromosome, every cell of the new conceptus will be completely deficient in factor VIII. This situation would lead to hemophilia A, a very serious bleeding disorder. If this man survives to adulthood and has his own children with a partner who has two X chromosomes with normal factor VIII genes, he will produce sperm that carry either the X chromosome with the factor VIII mutation or the Y chromosome. This must be true because the X and Y chromosomes segregate in meiosis from one another, just as if they were homologous autosomes (to review this point see Figure 21, p. 71). If one of this man's Y chromosomes fertilizes his partners egg to produce a new baby, the factor VIII mutation will be lost from the family. A boy without factor VIII deficiency will be born. If a sperm carrying this man's mutation-bearing X chromosome reaches the egg, the baby will be female, having inherited one X chromosome through the sperm and the other from the egg. The female conceptus will therefore have one X chromosome with the mutated gene and one with the normal gene. Will this girl have hemophilia A?

The answer is to some degree a matter of luck. When each cell of this female conceptus, as an early embryo, chooses its X chromosomes for inactivation at random, most of its cells may choose the one with the normal factor VIII gene. Under these circumstances, the only active X chromosome in a majority of cells will be factor VIII deficient. This circumstance could lead to a blood clotting defect. If, on the other hand, the X chromosome with the mutated gene is inactivated in most cells, they will be perfectly normal in their factor VIII production. Because most cells will be of this type, the woman will not be likely to manifest signs of factor VIII deficiency. In the real world there is usually a mixture of cells with one or the other X chromosome inactivated, and most of the time, when mutations are present on the X chromosome, disease is not present. Usually there are enough cells with the normal X chromosome active to prevent problems.

What will happen when this woman carrying factor VIII deficiency on one of her X chromosomes has her own children? If she passes the X chromosome with the factor VIII mutation to the egg, then the child will either be like herself with respect to factor VIII deficiency—a female carrier of the mutation (when an X chromosome-bearing sperm fertilizes)—or will be a male child afflicted with hemophilia A (if a Y chromosome bearing sperm fertilizes). If the normal X chromosome is passed to the egg, then of course all of the children will not carry the mutation and not only will have normal factor VIII levels but will have no risk of passing the factor VIII mutation to their own children.

It should be apparent from this discussion that diseases resulting from mutation of genes on the X chromosome are much more likely to manifest in males, who have only a single active X chromosome. Females, on the other hand, have millions of cells with one X chromosome inactivated and millions of others with the other X chromosome inactivated. With respect to alleles on the X chromosome, the female is has a "salt-and-pepper" distribution of cells, if we think of the salt grains as cells with one X chromosome inactivated and the pepper flakes as cells with the other X chromosome inactivated. She is accordingly referred to as a genetic mosaic.

Mosaicism for active alleles on the X chromosomes can have visible manifestations. The tortoiseshell cat, for example, is a female that has two different alleles of a gene on the X chromosome that affects fur coloration. Cells with one of the X chromosome inactivated give rise to an amber-colored hair follicle, whereas brown hair follicles are produced when the other X chromosome is inactivated. The result is an animal with patches of fur of different colors, with each patch representing a cluster of cells with one or the other X chromosome inactivated. Figure 25 shows a female mouse with a mutation on the X chromosome that affects coat color, much as does the tortoiseshell mutation in cats. This gene is called Tabby and causes hair follicles to produce yellow hair. As this photograph shows, there are stripes of tabby and normal fur. These stripes indicate the distribution of hair follicle cells with one or the other X chromosome inactivated. Because this mosaic coat color cannot occur without the

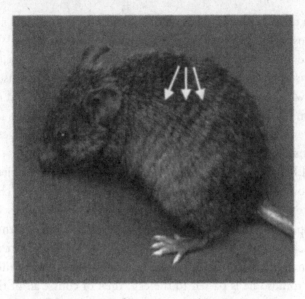

Figure 25   A living demonstration of the process of X chromosome inactivation. This female mouse carries a gene on the X chromosome that alters coat color (see text). When one or the other X chromosome is inactivated, groups of hair follicle cells, aligned in a striped pattern (arrows) exhibit a distinct coat color.

presence of two X chromosomes, these mice, and almost all tortoiseshell cats, are females.

Disorders of Chromosome Segregation

Given that tortoiseshell cats have their mottled coat color because of the patchwork distribution of cells with one or the other X chromosome inactivated, how could we ever have a male tortoiseshell cat? Such rarities occur because occasionally chromosomes do not segregate properly in meiosis. Consider the first meiotic division again (Figure 21, p. 71). Note how when one replicated chromosome segregates to one side of the cell, its counterpart goes the other way. But sometimes mistakes are made, and both pairs of an homologous chromosome go to one side and none goes to the other. This results in production of some eggs or sperm that have two copies of a chromosome instead of one and others that have no copies. These kinds of errors, collectively referred to as chromosome nondisjunctions, have very profound consequences. In older women, nondisjunction occurs more frequently as eggs proceed through meiosis, and after age 35 about 1 percent of eggs produced will have two copies of chromosome 21. When a sperm arrives with its copy, the conceptus has three copies instead of two and the fetus is born with Down syndrome. Because there are three chromosomes 21, the condition is referred to as trisomy 21. When the egg without any chromosomes 21 is fertilized by a sperm with a single copy of this chromosome, the result is a conceptus with a single chromosome 21. This is called, appropriately, monosomy 21, and live births do not occur in this situation. Trisomies and monosomies are generally very deleterious conditions and most of the time the result in death of the embryo or fetus. Trisomy for chromosomes 21 and 18, and for the X chromosome, are exceptions.

When meiotic nondisjunction involves the X chromosome, and the egg with two X chromosomes is fertilized by a Y sperm, the conceptus is XXY with respect to its sex chromosomes. As it turns out, the genes on the Y that determine that the fetus will develop as a male are dominant (see p. 82 for a discussion of dominant genes) and thus the conceptus will develop as a male despite having two X chromosomes. X chromosome inactivation will still occur, and this is how tortoiseshell male cats are occasionally produced. Nondisjunction of the X chromosome occurs, and the two X chromosomes have different alleles of the gene that affects coat color. This causes the mottled color of the tortoiseshell cat. Thus to produce a tortoiseshell male cat we must have allelic variation of X-linked genes that affect coat color while at the same time having a nondisjunction event that leads to an XXY male. Naturally, having all of these events take place on one conceptus is exceedingly rare, thus the high price commanded by tortoiseshell male cats. Although the best-recognized nondisjunction-related conditions involve nondisjunction in meiosis, such errors of chromosome segregation can occur whenever chromosomes segregate, whether that be in the first meiotic division, the second meiotic division, or even in mitotic chromosome segregation.

5

DEVELOPMENTAL BIOLOGY

Question: How could a woman give birth to her own identical twin?

Traditionally, developmental biology has been taught as a descriptive science in which is illustrated the emergence of new tissues, movement of cells, etc. Embryology texts that depict these processes are exceedingly complex; as such, they are difficult for an untrained reader to fathom. Fortunately, the incredibly intricate mechanisms of morphogenesis need not be understood at all to master the principles of development that are important to attainment of a clear view of germ line genetic engineering. If we look at development from the perspective of would-be gene manipulators, we need to understand a few basic things. These "themes" can be understood without any discussion of the details of cell multiplication, death, and movement that describe the creation of an embryo and fetus from a fertilized egg. These fundamentals will be discussed in the context of our goal of designing reasonable germ line gene manipulation strategies. There is, however, one period of development that must be discussed more specifically and that involves the phase that follows fertilization but precedes implantation of the embryo into the wall of the uterus, where it establishes contact with the mother's circulation via the creation of a placenta. Fortunately, this preimplantation period is short and easy to visualize, involves very few cells, and conforms to the principles of development that govern the far more intricate steps that take place after implantation. We will first discuss the broader thematic aspects of mammalian development and then take a closer look at the preimplantation period.

Examining Development From the Genetic Point of View

If we think about development from the fertilized egg to the adult with its 100,000,000,000,000 cells or so, we are faced with a logical conundrum. Once the sperm unites with the egg and the normal complement of genetic material found in all adult cells is established (that is, two copies of every gene), the remainder of development is, with few exceptions, accomplished by mitosis. The one-celled egg divides into two, two to four, etc., until the adult number is reached. Now remember that the purpose of mitosis is replication. The DNA is replicated as carefully as possible, and two identical copies of the genome are distributed to the two daughter cells when the parent cell divides (see Figure 20). Thus the adult is a colony of genetically identical cells—a genetic clone of the fertilized egg. If all cells at all stages of development have the identical complement of genes, how is tissue specialization, which is inevitably accompanied by specialized patterns of gene expression, accomplished? Development in mammals is of course subject to environmental influences, but these influences are lessened by the sequestration of the fetus behind the placenta and within the amniotic sac. To a large degree, then, prenatal development is autoregulated and relatively insensitive to outside factors. How then do cells that have the identical information available to them manage to achieve specialized patterns of gene expression that lead to and characterize differentiated cells? The answer to this question remains one of the challenging mysteries of developmental biology, and I would not be so presumptuous as to attempt to answer it here. However, when we review basic mechanisms of embryonic and fetal development, some of the factors that lead to differential gene expression will become apparent.

While on the subject of gene regulation, we should think of changes in this regulation over the course of organogenesis and cell specialization as a process of shutting genes off. Why so? Because the fertilized egg must, logically, be potentially capable of activating all of its genes. If this were not the case, it would be unable to give rise to all the specialized cells of the adult, which perforce exhibit as a group the totality of patterns of gene expression available to the organism. So, as the egg divides the housekeeping genes that keep the cells alive remain active in all of its mitotic descendants; but as tissues become increasingly specialized the repertoire of expression of the other genes becomes progressively more limited. That is, once a cell becomes a part of the heart, it must be able to access genes that produce contractile proteins of heart muscle but it need not be able to access genes that are specific for other specialized tissues like liver, skin, or kidney. In fact, inappropriate activation of the wrong tissue-specific genes can be associated with serious disease. Therefore, it is reasonable to state not only that genes specific for tissues other than those occupied by any given cell *need* not be active but, further, that they *must* not be active. Therefore, we may regard development as a process of progressive limitation of genes available for expression to each cell, with fewer and fewer genes remaining available for activation as development proceeds. Development is a process of shutting genes off just as much as it is a process of turning genes on. It is important to keep this principle in mind when we turn our attention to genetic engineering.

Determination Versus Differentiation

Another important principle of embryonic and fetal development that is intuitively obvious but also requires mention is the distinction between determination and differentiation. Determination is a process whereby cells are assigned to a specific developmental program but have yet to carry out that program. It is easy to appreciate that the cells which function in our livers must have at some point in fetal development been designated to become liver cells without having yet established themselves as differentiated liver cells, with liver-specific patterns of gene expression. How do cells with multiple possible developmental choices get committed to specific patterns of gene expression? Not surprisingly, there are genes that effect such commitments.

Induction

Induction is the process whereby interactions between cells trigger a pattern of gene expression that steers the responding cell down a specific path of differentiation. Induction can be through direct contact or can result from stimuli released from distant cells that reach the target cell via the circulation. Examples of this phenomenon illustrate the importance of cell interactions during development.

Induction by direct cell contact is a phenomenon best illustrated by experiments with amphibians—frogs and newts—because these creatures develop outside of the body of the mother where they can be readily accessed for experimental manipulations. Newts can have side appendages called "balancers," which protrude from the flank of the animal and function much like the pole used by a tightrope walker—they keep the animal upright and prevent it from rolling on it side. Some newt species have these balancers, and some do not. If skin from a species that does not have balancers is grafted to a region of another species where balancers normally form and the corresponding skin from the graft recipient's flank is removed, the grafted tissue will form a balancer, even though the species from which it was obtained does not have this structure. Clearly, contact with the underlying tissue of the graft recipient results in a pattern of gene expression normally not undertaken at all by cells from the donor species. If the graft is made to a region of the recipient where balancers do not form, the donor tissue does not form a balancer, a finding that demonstrates that the induction effect is localized and caused by direct cell contact. Interestingly, if similar experiments are done at a later stage of embryonic development, when the pattern of gene expression of the donor tissue is already restricted, the induction is not successful. These findings show that the response to inductive stimuli is dependent on the responding cells' ability to activate the appropriate genes, and they reinforce the notion that as development proceeds, the repertoire of gene expression becomes progressively more restricted.

The ability for induction to take place at a distance is very well illustrated by a naturally occurring genetic abnormality that occurs in several species of mammals and affects development of external sexual characteristics. As noted in our discussion of genetic inheritance, the Y chromosome contains a gene that induces the formation of testicles and this gene is dominant (see pp. 82 and 87). Once the testis is formed, it produces two important hormones, testosterone and another hormone that blocks the development of fallopian tubes and a uterus. This second hormone is called müllerian duct inhibiting factor, or MIF. Testosterone works by crossing the cell membrane, where it is captured by a protein and taken to the cell nucleus, where it alters the pattern of gene expression. The protein that captures testosterone in the cytoplasm is called the androgen receptor. Mutations can occur in the androgen receptor gene that totally obliterate receptor function and render the cell insensitive to the effects of testosterone. Under these circumstances testosterone can enter the cell, but it can have no effect because it cannot reach the nucleus. Some people who have this genetic abnormality have the XY chromosomal complement found in a normal male. However, because testosterone produced by their testicles cannot function, the development of male external sex organs does not take place and these individuals have the external appearance of normal females. The testicles in these women still produce the other hormone,

MIF, which blocks the formation of a uterus and fallopian tubes. The result is a woman with testicles in the abdomen and no uterus or fallopian tubes. These women often have very pronounced female characteristics, because the small amount of the female hormone estrogen in their bodies is totally unopposed by the action of testosterone. In addition, they have very little body or facial hair, the growth of which is stimulated by testosterone. The problem is usually discovered when the developing girl fails to have menstrual periods, which of course cannot take place in the absence of a uterus. This genetic abnormality is known as testicular feminization, or Tfm for short.

The Tfm mutation shows the inductive effects of testosterone at tissues physically distant from the testicle are unable to develop. It also shows the negative inductive effect of MIF, which blocks physically distant tissues that normally give rise to a uterus and tubes from differentiating into these structures.

When considering the prospect of genetic engineering, it is important to appreciate the complex interactions of cells, either from direct contact or distant communication, to the development of organ systems. When developmental problems arise, they could be due to failure of an inducing cell to elaborate the appropriate stimulus but they could also be due to a failure to respond properly to an inductive stimulus.

Pleiotropism

Pleiotropism is a term used to explain another important principle of development. An understanding of this principle can be attained by thinking of development as the sprouting of a tree. When the embryo has very few cells, there are only two branches on this tree: One branch leads cells down that pathway of development into placenta, amniotic sac, and other supporting tissues for fetal development, and the other designates cells to become the embryo itself. However, shortly after implantation of the embryo into the uterine wall, its cells are allocated into three major pathways of organogenesis. Some become skin, some become tissues like muscle, bone, and blood, and some become visceral organs such as liver. These branches are called ectoderm, mesoderm, and endoderm, respectively. We may now consider the developmental tree of the embryo to have three main branches. As these branches sprout further more limited and specialized pathways of cell differentiation emerge, and these can be thought of as smaller branches of the tree. Finally, when the major organs have arisen and cells are either carrying out their highly specialized tasks or fully committed to pathways of differentiation that lead to execution of those tasks, the branches of the tree are complete. We may regard birth as a landmark for essential completion of this developmental arborization. Now, of course, there are inductive stimuli that send cells down one pathway or another. When mistakes occur in development, they may be the result of a failed induction due to a defect in the inducing cell or a failure to respond appropriately to an inductive stimulus. Let us now envision cells at a branch point in development, with one of the cells, or groups of cells, inducing a pathway of differentiation in its target cell but then proceeding itself down a separate pathway. If this inducer cell has a defect, the responder cells may not differentiate properly. However, this defect may never affect the ability of the inducing cell to complete its own program of differentiation. Another possibility is that the defect in the inducing cell may manifest much later in development as an abnormality in its own differentiation.

The result of these kinds of errors is that a single genetic defect can have effects on many different, seemingly unrelated tissues. These kinds of effects are termed "pleiotropic effects." An example in mice is a gene called W, for white spotting. This is a semidominant mutation (for a review of semidominant effects see p. 82) that, when present on both chromosomes, causes the animal to have no sperm or eggs, no hair pigment, and a severe anemia, with very few blood-forming cells. This pleiotropic mutation is known to be due to the absence of a single inducer molecule called *ckit*. Interestingly, there is another semidominant mutation called Steel, or Sl for short, which has precisely the same effects but which is clearly a distinct gene from W. Molecular analysis has shown that in the Sl mutation, the cellular receptor for *ckit* is defective and nonfunctional. In the W mutation, conditions such as anemia can be corrected by bone marrow transplantation with cells that do not have the mutation. This shows that the bone marrow matrix in W animals provides an environment that can support the formation of blood cells and that in W mice it is the blood cells that are defective. However, bone marrow transplantation is completely ineffective in Sl, a finding that forces the conclusion that the bone marrow environment is defective in this mutation. As you might predict, it is possible to rescue W mice from fatal anemia by transplanting Sl bone marrow cells, if enough of these cells can be isolated for transplant.

Pleiotropism is a phenomenon that should be kept in mind when considering germ line genetic intervention. It is possible that alterations of gene function engineered into the early embryo could have quite unexpected effects on tissues thought not to be affected at all by the gene of interest. For this reason, as we will discuss in more detail later, a very thorough understanding of the action of genes one intends to modify should be in hand before any modification of the human genome is undertaken. Another important point is that environmental insults that damage cells can cause quite a diversity of developmental abnormalities in the newborn. If you understand the principle of the branching tree of development, you should be able to predict that insults to the developing fetus, be they genetic or environmental in origin, are more likely to have protean manifestations if they occur relatively early, when the main branches of the tree are just forming.

Early Development

As I mentioned at the beginning of this chapter, the intricate cellular movements and interactions that characterize fetal development need not be known to meaningfully evaluate genetic engineering strategies. However, the very earliest phases of embryo development—those that take place within a few days of fertilization and that precede implantation of the embryo into the uterine wall—should be understood, because it is in this phase of development that manipulations intended to modify the genome are performed.

Let us first consider fertilization—the union of sperm and egg. From the point of view of both sexes, this is an exceedingly inefficient process. Girls are born with about 400,000 eggs in their ovaries, yet only about 400 of these will be released and have the opportunity to be fertilized during their lifetimes. However, this inefficiency is nothing compared to that in males. Men may release as many as 100 million sperm into the female reproductive tract at the time of intercourse, yet only a few hundred of these ever reach the egg and only a single one is able to fertilize. Moreover, whether or not a male is sexually active, he continues to produce millions of sperm on a constant basis throughout his adult life. If a man has three children, he has used three out of billions of sperm produced from the testes during his lifetime.

In humans, of course, mating is not linked to egg release, or ovulation, as it is in many mammals. Human females are willing to mate at any time in their cycle of ovulation, and they may in fact deliberately attempt to avoid intercourse when they think ovulation is taking place. This, of course, is the rhythm method of contraception. The shocking inefficiency of fertilization is relevant to genetic modification of the germ line for the reason that attempts to effect such modification by delivering genetic material to the testes or ovaries is not likely to succeed. Most of the time, the gametes that receive new genes will not be those that participate in fertilization.

Now we all know that chicken eggs have a shell that must be broken if breakfast is to be eaten; but it may be surprising to know that the human egg also has a shell. Although the shell of a bird egg is hard and opaque, the shell of the mammalian egg is soft and clear, so that one can easily see the egg cell, or oocyte, right through the shell. Because this shell looks like a clear zone surrounding the egg, it is referred to as the zona pellucida. The zona, as we'll call it for short, is laid down around the egg as it proceeds toward ovulation and progresses through meiosis. When the egg is released the zona can be easily seen surrounding the egg, and often the first polar body, released after the first meiotic division, can be seen tucked under the zona (for a review of meiosis in the female and polar body release, see pp. 74–75). The eggs develop within chambers, or follicles, of the ovary. Figure 26 shows a developing egg within a follicle. Note that the follicle is well separated from the surrounding ovary by a wall, the follicle cells both line the cavity of the follicle and surround the egg, and the follicle has fluid inside. This fluid accumulates until the follicle ruptures at ovulation. In this figure you can also see the DNA stained within the nucleus. At the stage where this photograph was taken, DNA has replicated and crossing over is complete.

Follicle cells surrounding the egg may still be found attached to the zona after ovulation. To fertilize, the sperm must penetrate the halo of follicle cells and then must traverse the zona pellucida. Why place so many obstacles to

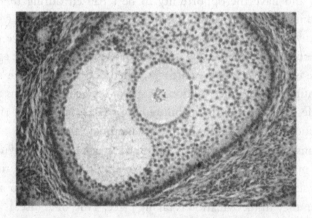

Figure 26 An egg developing within a follicle. Note the follicle wall and the substantial number of small follicle cells that surround the egg. The chromosomes, condensed and having just completed crossing over, are seen stained within the nucleus.

sperm pentration, when fertilization is an absolute prerequisite for procreation, which in turn allows the organism to complete its best effort at replication and sustain the species as well?

There are two very good reasons: First, although it is important for sperm to reach the egg, it is also crucial that only one sperm participate in fertilization. Sperm bring one copy of the human genome to the conceptus, with the egg supplying the other copy. If two sperm were to enter the egg, three copies of the genetic material would be present, and the abnormally high levels of gene expression would kill the embryo after just a few days of development. The follicle cells, and particularly the zona pellucida, help prevent the disaster of polyspermic fertilization. Second, when the one-cell embryo divides into two and then four cells, and is free floating in the fallopian tube, the risk that these cells will not stay closely associated is relatively great. The zona keeps the cells contained snugly in a nice small chamber where they won't lose contact with one another.

How do sperm get through the layers of cells and zona matrix that surround the egg? The follicle cells are rather easily penetrated. They are loosely connected with a gluelike substance called hyaluronic acid, and the vigorously motile sperm get between them fairly well to reach the zona. But what then? The zona pellucida actually helps the sperm reach the egg. To penetrate the zona, the sperm must first attach to it. If sperm simply "bounce off" the zona, fertilization will never take place. To prevent this mishap, the zona actually contains proteins that avidly bind to the surface of the sperm. The next step in zona penetration involves the boring action of the highly motile sperm but almost certainly also involves digestion of a small track in the zona by enzymes that surround the head of the sperm. The release of these sperm enzymes is linked to the binding process. To see how these events are linked we need to take a closer look at the sperm.

The nucleus of the sperm is surrounded by a bag of digestive enzymes called the acrosome that is much like a giant lysosome (see Figure 1). It is easy to see how the sperm head with its nucleus, acrosome, and surrounding plasma membrane is put together as follows: Make a tight fist and plunge it into a very soft pillow. Then push your fist and surrounding pillow into a large plastic bag. Now imagine that you can remove your arm at the shoulder, place it in the bag, and seal the bag. In this model your fist is the sperm nucleus, and the skin around it is the nuclear membrane. The pillow is the acrosome. The part of the pillow that makes contact with your skin is the inner membrane of the acrosome. The outside surface of the pillow, which makes contact with the overlying plastic bag, is the outer acrosomal membrane. The contents of the pillow represent the enzymes within the acrosome, and the overlying plastic bag represents the outer membrane of the sperm cell.

When the sperm binds to the zona, the protein in the zona triggers a reaction that is equivalent in our model to taking a knife and slicing through the plastic bag and through the underlying pillow surface, then sewing the cut edge of the plastic bag to the cut surface of the pillow immediately below. This, of course, opens a gaping hole in the pillow, and the feathers spill out. When the sperm contacts the zona, multiple such cuts, followed by fusion of the plasma membrane with the outer acrosomal membrane immediately beneath it, occur in several places, and the contents of the acrosome are released just at the site where they are needed to help the sperm digest its way through the zona—that is, at the binding site of sperm and zona. This process of disrupting the acrosome by fusing its outer membrane with the plasma membrane is called the acrosome reaction, or AR. It should be clear that the plasma membrane, our plastic bag, is also disrupted during the process of sperm penetration.

The process of penetrating the zona is fomented by the zona proteins that facilitate sperm binding and induce the AR but is also a somewhat time-consuming step that tends to stagger arrival of sperm to the egg surface and discourage binding of multiple sperm to the egg surface. Once a sperm gets past the zona and fuses with the egg, the egg has further responses that help to block entry of additional sperm. Small granules, called cortical granules, sit immediately under the egg plasma membrane, and entry of the sperm cell triggers fusion of these granules with the egg plasma membrane and release of the granule contents into the space between the zona pellucida and the egg. Cortical granules contain enzymes that, when released, drift across the space between the egg surface and zona and partially digest the proteins of the zona. This digestion alters the structure of the zona proteins so that they are no longer able to bind sperm. This mechanism discourages penetration of multiple sperm to the egg surface. In addition, contact with the sperm causes changes in the egg membrane that prevent fusion of additional sperm that might have made it through the zona before the cortical granules were released. Thus the egg and its surrounding zona both encourage sperm binding and penetration, while discouraging polyspermic fertilization. Obviously, the steps taken to block polyspermy take some time, and occasionally more than one sperm will enter the egg before the block is in place. These embryos, with three copies of every gene instead of two, are doomed to die.

After the sperm enters the egg, changes occur that confer upon the fertilized egg an arrangement of the genetic material that is unlike any other cell. As you know from our model of the typical mammalian cell (see Figure 1), the genetic material is sequestered within the nucleus. However, when the sperm arrives to the egg, its genetic material is contained within the sperm head and is not ad-mixed with the genetic material from the egg. In addition, at the time of fertil-ization, the egg is in the middle of meiosis and only completes the second mei-otic division, with release of the second polar body, after fertilization (see Figure 22, p. 75 for diagram of meiosis in the egg). As a result of these un-avoidable realities, the sperm and egg genetic material are physically separated in the hours immediately following fusion of the cells. The sperm head decon-denses to form a membrane-bound compartment that contains the father's copy of the genes, and the egg completes meiosis, releases the second polar body, and surrounds the chromosomes that remain inside within their own membrane. These two compartments are like nuclei but contain only half the amount of genetic material typically found in a nucleus. To distinguish them from a normal nucleus they are called pronuclei. Thus the fertilized egg has two pronuclei, a "male" one with the sperm chromosomes and a "female" one with the egg chromosomes. Pronuclei are shown diagrammatically in Figure 22, and Figure 27 shows a photograph of an unfertilized mouse egg sitting next

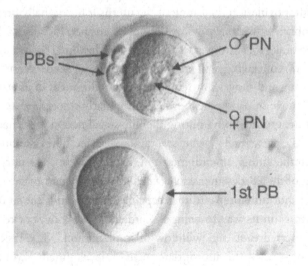

Figure 27   A fertilized (top) and unfertilized egg, shown for comparison. Note that the unfertilized egg has only a single polar body, whereas the fertilized egg has 2 polar bodies (see Figure 22 and text). The male and female "pronuclei" containing the fa-ther's and mother's genes are shown.

to a fertilized egg. Note that the unfertilized egg contains no membrane-bound structures inside and has a single polar body under the zona. However, the fertilized egg has two membrane-bound pronuclei inside the cell and has two polar bodies, indicative of the completion of meiosis by the egg. It's also easy to see that the cytoplasm of the fertilized egg is relatively granular in appearance, a difference that reflects the cytoplasmic changes that occur after sperm-egg fusion. Another feature that is difficult to miss is the presence of spherical structures within the pronuclei. These objects are clusters of genes that produce the RNA used to make ribosomes, the mRNA translation machines (see Figure 1 and p. 35). It is indeed amazing that one can look at a fertilized egg under a low-power microscope and see a cluster of genes! Shortly after fertilization the pronuclei begin replicating their DNA in preparation for the first mitotic division of embryonic development. In some species the pronuclei fuse to make a single nucleus before the first division, whereas in others they break down as the chromosomes line up in the equator of the cell, attach to the spindle fibers, and prepare to split at the centromere for mitosis.

There are many remarkable features of the embryo's development within the first few days after implantation that, although striking and unpredicted, make a lot of sense when one considers the tasks that the embryo faces. The female reproductive cycle is designed to maximize the chances of conception. Therefore, after ovulation the lining of the uterus is hormonally prepared to receive the embryo (this subject will be discussed in more detail in Chapter 6 on reproductive biology). However, if the egg is not fertilized for whatever reason, the chances of conceiving will be reduced unless, at some point, the female gives up waiting and moves on to initiate the development of another egg. The decision to "move on to the next cycle" is popularly recognized as the physical shedding of the uterine lining during the menstrual period, but changes indicative of a failed cycle actually occur several days earlier. The take-home message is that the uterine lining, or endometrium, is receptive to the embryo for only a few days out of the 28-day menstrual cycle. In the early part of the cycle the endometrium is not fully prepared to accept the embryo, and late in the cycle the endometrium is on its way to being discarded. The few days of receptivity are commonly referred to as the "window of implantation." The task of the embryo, then, is to reach the appropriate stage of development and be present in the uterine cavity sometime within the window of implantation. Failure to implant is the ultimate disaster for any new conceptus, and thus evolution has selected for a strategy of preimplantation development that maximizes the opportunity for implantation.

Preimplantation Embryos Develop Without Growing

The first surprising feature of preimplantation development is that cells divide without any growth. When most cells divide they have two growth periods, one before mitosis designed to muster the resources needed for DNA replication, and one after mitosis to restore the size of the daughter cells to that of the original parent. However, the division of cells in the embryo immediately after fertilization is characterized by no growth but, instead, simple "cleavage" of the cells, with the cell size dropping by 50% with each successive mitosis. Figure 28 shows a group of mouse embryos that demonstrate this phenomenon. Shown are a 1-celled embryo with its pronuclei, a 2-celled embryo, an 8- to 16-celled embryo, and a blastocyst, which is a hollow ball of about 32 cells. As you can see, the embryo occupies the same space within the zona pellucida as development proceeds, even though cell numbers are increasing, and cell sizes become progressively reduced. This latter point is most easily appreciated by comparing the two- and one-celled embryos. The best way to understand this unusual behavior is to appreciate that the egg is a very large cell and has enough material for many smaller cells. The accumulation of resources during egg de-

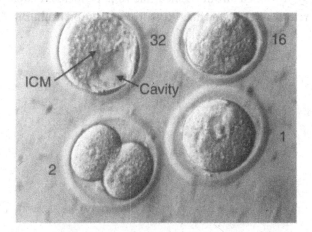

Figure 28 Mouse preimplantation embryos at various stages of development. The approximate number of cells at each stage is shown to the right. Note that no growth occurs even though cell numbers increase from 1 to about 32. The ICM (upper left) contains cells that will become the embryo, whereas the cavity of the blastocyst is lined with cells that will produce supporting tissues for embryonic development such as the amniotic sac.

velopment, or oogenesis, allows the cells to divide after fertilization without de-
lays that could cause the embryo to miss the window of implantation. While
we're looking at Figure 28, there are a couple of additional points to touch
upon that we will return to later. These relate to the two more advanced embryos. First, it is easy to appreciate that, although the cells of the 2-celled embryo are clearly distinguishable from one another, those of the 16-cell embryo
are closely compacted such that their demarcating plasma membranes are difficult to see. This compaction results from tight seals that form between the
cells, thus preventing leakage of the fluid that accumulates in the cavity of the
blastocyst, shown to the immediate left. A couple of features of the blastocyst
are also important to note. First, it is easy to see that the cells around the fluid-
filled cavity are not evenly distributed. Rather, there is a mass of cells at one
end called the inner cell mass, labeled in Figure 28 as the ICM. The ICM is the
region from which the fetus develops, while the cells lining the fluid-filled cavity later produce extraembryonic tissues such as the amniotic sac. We will discuss the ICM in more detail later. Second, the blastocyst is the last stage of development before attachment of the embryo to the uterus for formation of a
placenta and more advanced fetal development. If implantation does not occur
shortly after blastocyst formation, the embryo will die. To implant, the blastocyst will have to "hatch" from its shell, the zona pellucida. Hatching is accomplished by release of enzymes that digest the zona proteins but is also probably

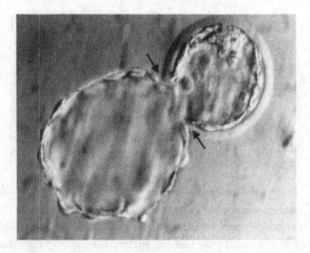

Figure 29   A hatching blastocyst. The points at which the zona pellucida is ruptured
are shown with arrows.

helped along by pulsatile expansion and contraction, which helps the embryo push its way out. A blastocyst in the process of hatching is shown in Figure 29.

Preimplantation Development is Remarkably Plastic

Of course, rapid development is not the only imperative faced by an embryo whose genotype will be lost for eternity if the window of implantation is missed. What if an accident occurs and one of the cells of the embryo dies during cleavage? Compensatory mechanisms to deal with such contingencies also exist and are manifest as a plasticity of development that is truly startling.

We can create an experimental situation that mimics loss of half of the cells of the preimplantation embryo: All we have to do is remove the zona pellucida of a two-celled embryo and give the cells a shake. If we are persistent and gentle, we can separate the two cells and culture them individually. How does the embryo respond to this manipulation? Figure 30 gives the answer. Figure 30A shows a two-celled embryo that is intact and another in which the two cells have been separated. The two separate cells divide just as if nothing happened (Figure 30B), and produce "miniblastocysts" at precisely the same time as a normal embryo. Figure 30C shows two such mouse "miniblastocysts," photographed alongside an embryo that had the zona removed but was otherwise not manipulated. As you can see, the unmanipulated embryo has produced a blastocyst twice the size of the two embryos that developed after cell separation. What we learn from this simple experiment is how the embryo compensates for loss of up to half of its cells. Cell division cannot be sped up to replace the lost cells, because it is already taking place at maximum speed. The embryo cannot afford to take the time to replace the cells, for to do so would delay formation of the blastocyst and cause the embryo to miss the window of implantation. The embryo therefore compensates for the loss by simply pretending that nothing happened and producing blastocysts with half the number of cells that are fully capable of developing normally. In this example the two small blastocysts are of course genetically identical, and if they were both returned to a female for implantation they could develop into identical twins. The ability to produce genetically identical twins or clones by embryo splitting is exploited in the cattle industry when a breeder obtains an embryo believed to carry a mix of genes that will produce a valuable adult animal. To produce two adults instead of one, the embryo is simply cut in half with a knife and the two halves are implanted. This procedure can be done right on the farm by flushing embryos

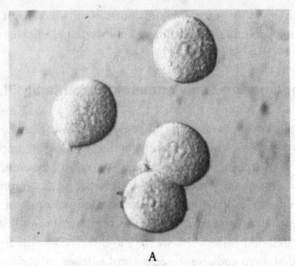

A

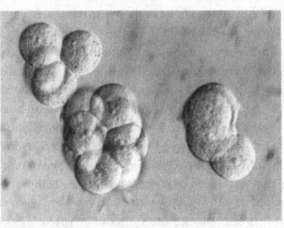

B

Figure 30   Cells separated at the two-cell stage develop normally and at the normal rate. A. A two-celled embryo immediately after cell separation, with an intact embryo shown for comparison. B. The same cells as in A after about 24 h in culture. Note that the cells are all the same size but the number of cells in the embryos that developed after cell separation is half that of the unmanipulated embryo.

from a female, cutting them under a microscope, and returning them to a female for continued development.

Once might ask what happens if the same separation of cells is performed at the four-cell stage. When this is done very tiny identical quadruplet blastocysts are formed, and, again, they have 25% of the normal number of cells in a blas-

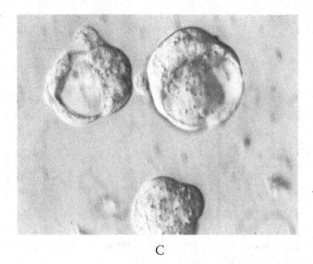

C

Figure 30   C. A normal blastocyst and two "mini-blastocysts" resulting from the embryos shown in A. A cavity has not yet begun to form in one of the mini-blastocysts.

tocyst and can develop into quadruplet offspring that are genetically identical. When separations are performed after the four-cell stage, however, the embryo runs into a problem: When it attempts to make a blastocyst, it no longer has enough cells to produce both the cell-lined cavity and the inner cell mass (ICM). Faced with this crisis, the embryo chooses to attempt formation of the blastocyst cavity, but of course, without an ICM, these clusters of cells cannot produce a fetus after transfer to the uterus.

When cells are removed from the preimplantation embryo it continues to develop to a blastocyst with fewer cells. What happens if we add cells? Remarkably, the embryo welcomes additional cells and incorporates them, developing into a blastocyst with more than the normal number of cells. Figure 31 shows an example of this behavior. In this experiment, the zona pellucida was removed from two mouse embryos at the eight-cell stage and the embryos were then simply pushed together. When this is done, the two separate embryos become one and produce a giant blastocyst as shown in the figure. This blastocyst, when transferred to the uterus, will implant and develop into a normal-sized offspring, indicating that the embryo compensates at some point for the excess cells and restores the normal cell numbers to the fetus. It's easy to imagine what happens when embryos are disaggregated and implanted as small, genetically identical blastocysts—you get twins or even quadruplets. But when embryos are aggregated the result is not so easy to predict, because now a single

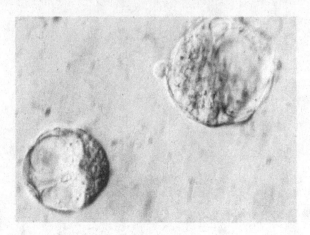

Figure 31 A giant blastocyst (upper right) resulting from aggregation of two embryos, shown with a normal blastocyst (lower left).

embryo is composed of two genetically distinct cell types. Figure 32 shows what happens when an embryo with the genes for white fur (albino) is aggregated with another embryo that carries the gene for black coat color. Notice that the resultant animal is a patchwork, or mosaic, with groups of cells from each contributing embryo easily recognized as patches of white or black fur. A mottled coat color does not represent a form of genetic mosaicism that is likely

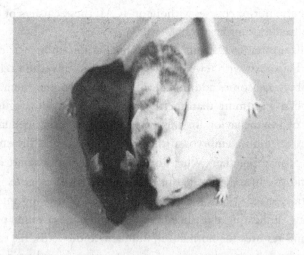

Figure 32 An adult, genetically mosaic mouse born after aggregation of embryos carrying genes for albino fur (mouse on the right) and black fur (mouse on the left). Note that the mosaic contains patches of cells derived from each of the donor embryos.

to affect the animal's health or well-being, but consider what might happen if the embryo with the white fur had XY sex chromosomes and the one with the black fur was XX? When male and female cells are developing in the same animal confusion could certainly result, and it sometimes does. However, most of the time these "sex mosaics" become males, producing sperm derived only from the XY component of the animal. Excited by the amazing ability of separate embryos to communicate with each other and cooperate to form a single composite embryo, scientists have experimented with even more challenging situations than mosaicism for the sex chromosomes. Rat embryos have been aggregated with mouse embryos, and although it is possible for these interspecies mosaics to implant in the uterus of a mouse, the rat cells still behave as if they're trying to produce a rat, which is about 10 times the size of a mouse. They consequently overgrow the embryo and destroy it. When closely related species of similar size are subjected to embryo aggregation, however, it is possible to obtain live animals with cells from two different species. The most spectacular example is the "geep," an animal produced by aggregating embryos of a goat and a sheep. It is quite a strange-looking creature.

The ability of the embryo to accept additional cells extends to the blastocyst stage. If ICM or similar cells are injected into the fluid cavity of the blastocyst, they can crawl into the ICM and contribute to the resultant fetus and newborn.

The First Step of Determination Takes Place Before Implantation

In our previous discussion of development (see p. 91) we distinguished determination, a process whereby cells are assigned to differentiate into specialized types, from differentiation, the process of actually becoming specialized. The blastocyst is an example of cells that have taken two different developmental branch points. Those cells around the cavity are destined to become the amniotic sac and other extraembryonic membranes, whereas cells in the ICM become the embryo proper. The conclusion is therefore inescapable that the cellular interactions leading to "commitment" of cells, and the associated changes in gene regulation, occur before formation of the blastocyst, at a time when the embryo can be removed to the laboratory for study. For this reason, preimplantation development presents a fascinating opportunity to study determination.

In a developing ball of genetically identical cells, how is it decided which ones will occupy the ICM? Experiments indicate that the greater the amount of surface contact a cell has with its neighbors, the greater the chance that the cell will enter the ICM. If cells from two genetically distinct eight-celled embryos (like those contributing to the mosaic mouse in Figure 31) are disaggregated and reassembled such that a cell with the albino (white coat color) genes is surrounded with cells carrying the black coat color genes, a blastocyst will develop that, when transferred, will lead to birth of a pup that is predominantly albino. This result indicates that the cell in the "middle of the ball" is more likely to enter the ICM. Another experiment that reinforces this conclusion consists of repeated separation of cells after each cleavage division. This procedure leads to cells sitting in isolation, without any surface contact with other cells. When the time comes to make a blastocyst, these isolated cells simply fill up with the fluid that is normally founding the blastocyst cavity and no cells attempt to become ICM. In the context of our previous discussion of induction by direct contact, we may speculate that simple contact between cells is sufficient to induce determination of the ICM.

Considering Early Development from the Point of View of Gene Regulation

As suggested in our foregoing discussion of gene regulation during development, determination can be thought of in another way—as a restriction of the gene expression repertoire. Logically, the fertilized egg must have the full repertoire available, because it must give rise to all of the cells that emerge as development progresses. However, once cells have become allocated to the extraembryonic portion of the early embryo, and line the blastocyst cavity, they are far more restricted with regard to the genes they are able to express. We know from construction of identical twins and quadruplets by embryo splitting that each cell in the two- and four- celled embryo still retains the ability to activate any subgroup of genes needed for function as any specialized tissue. The separation of the ICM and the other cells in the blastocyst therefore represents a key step in which some cells of the embryo lose their developmental totipotency. As we examine the blastocyst (Figure 30) we appreciate that the cells of the ICM have not lost that totipotency. They still can give rise to cells of the amniotic sac and placenta, but for ICM cells these patterns of differentiation wait for the next generation before they appear. However, for procreation to continue, these ICM cells must eventually produce extraembryonic tissues and thus must retain the full gene expression repertoire.

Gametes and Early Embryos Can be
Frozen and Stored Indefinitely

We have all heard stories of individuals diagnosed with terminal illness who have themselves frozen so they may be thawed after a cure for their malady is found. Freezing of adults is not technologically feasible because of one mole-cule—water. When water is frozen it forms crystals that lacerate cells and rupture them. Were an attempt made to remove sufficient amounts of the water from an adult human to avoid these effects when temperatures were lowered, death would certainly result. However sperm, eggs, and embryos can have substantial amounts of water removed without dying. This is particularly true of sperm, which are very condensed cells with a relatively low water content. In addition, killing of even half of the sperm cells in a sample would still leave millions alive to fertilize eggs either by artificial insemination or in the tissue culture dish. Embryos are also fairly easily frozen. Even though the cells are large, survival is good after desication and freezing, and when the embryo cleaves several times before freezing the cells are smaller and more numerous. Eggs are rather difficult to freeze, although they have been successfully frozen. Once sperm, eggs, or early embryos are successfully frozen they can be stored more or less indefinitely. Embryos have been frozen and subjected to an amount of irradiation equivalent to several hundred years' worth of back-ground cosmic radiation that we all experience on a constant basis. When thawed and transferred, these embryos developed no differently from those that are frozen for only a few hours.

How Can a Woman Give Birth to Her Own Identical Twin?

The ability to freeze the cells involved in the earliest phases of procreation is of enormous importance to the issue of germ line gene manipulation, because freezing makes possible many strategies for such manipulation that would oth-erwise be unworkable. As an example, consider the question posed at the head of this chapter. From the material we've just reviewed, it should be easy to pro-pose a strategy for allowing a woman to give birth to her own identical twin. To accomplish this feat, it is necessary only to cut an XX embryo in half, freeze one half, and implant the other half. When the embryo gives rise to a baby girl and she grows to adulthood, we can then retrieve the other half-embryo from the freezer and transfer it to her. If and when she becomes pregnant and delivers, the child will be genetically identical to her, having been derived from the same conceptus. Perhaps in this case the woman should consider her daughter as representing her "better half"! To have a woman give birth to her own identical twin is in fact a successful cloning procedure, and the technology available to accomplish cloning by this approach is already in hand. The more elaborate methods of cloning that have stirred so much controversy will be described in more detail later. These latter approaches are not currently workable but are more controversial because they might make possible the creation of hundreds of genetically identical individuals. Cloning by embryo splitting is limited essentially to the production of quadruplets by the fact that sufficient cell numbers must be available after embryo splitting for the embryos to make a blastocyst. Of course, this is not to say that cloning by embryo splitting would go unnoticed. The birth of identical twins to separate mothers, which could result if two half-embryos were implanted into separate women, would certainly stimulate a vigorous debate on the ethics of embryo manipulation. We will visit these discussions once a bit more of the china and silver is in place.

6

REPRODUCTIVE BIOLOGY AND ASSISTED REPRODUCTIVE TECHNOLOGIES

A man who has had a vasectomy marries a woman who has had a hysterectomy, and they decide to have children but will not adopt for religious reasons. One year later they are the proud parents of a baby boy and girl. Question: How did they do it?

In this chapter we will take a focused look at aspects of reproductive biology and assisted reproductive technologies that have relevance to engineering of the human germ line. To appreciate both the limitations imposed and the opportunities offered by the processes of reproduction for genetic manipulation of gametes and embryos, it will be necessary to examine the development and differentiation of sperm and eggs and the mechanisms controlling release of these cells from the testicles and ovaries, after which they ultimately unite to form the new conceptus. Once these processes are understood, it will be easy to see how medical intervention can exploit them for genetic engineering purposes.

To trace the origin of sperm and eggs, we can return to the ICM of the blastocyst. Recall that within the ICM reside cells that retain developmental totipotency. Shortly after the blastocyst implants into the wall of the uterus

The Science and Ethics of Engineering the Human Germ Line: Mendel's Maze, by Jon W. Gordon
ISBN 0-471-20647-4 Copyright © 2003 John Wiley & Sons, Inc.

and begins to form both the embryo and the extraembryonic membranes, a few cells from the ICM are set aside, or determined, to become germ cells. These cells still divide only by mitosis and are genetically identical to all other cells in the embryo. Shortly after these cells arise, they begin to crawl like an ameba to a region up under the developing spine, and they then move laterally to occupy regions known as the genital ridges, which will become the future gonads. During the migration period the cells also divide rapidly, such that they number in the thousands by the time the genital ridge is reached. At this stage these cells are referred to as primordial germ cells. In the female these cells divide to reach several million in number and then stop. However, many then degenerate as the fetus develops, such that when a baby girl is born, the ovaries will have about 400,000 of these cells. In females the germ cells become surrounded by tightly apposed cells of the ovary and immediately begin the process of meiosis. That is, they replicate their DNA, their chromosomes line up next to one another, and crossing over occurs (see Figure 26, Chapter 5). After crossing over, however, meiosis stops, and the chromosomes remain associated. These eggs, surrounded by a single tight layer of follicle cells, are called primordial follicles. Eggs remain arrested in this state unless or until they are stimulated to ovulate after puberty is reached. Over the life of a woman, about 400 eggs out of the 400,000 in the ovary are ovulated. In men, the primordial germ cells arrive at the genital ridge and then remain dormant until puberty.

The best way to understand the process whereby germ cells are stimulated to complete meiosis and differentiation and prepare themselves for release from the gonads is to consider the process from the perspective of the brain. From birth until puberty, the brain is completely uninterested in the levels of sex hormones—testosterone in men and estrogen in women. However, puberty is characterized by a developmental change in the brain that results in a new-found interest in sex hormone levels, which, before puberty, are low. In boys the brain begins to demand the production of testosterone from the testicles. To obtain this production, the brain releases hormones that travel to the testis and causes cells in the testis to produce testosterone. This change in the testis is accompanied by proliferation of the spermatogonia, as the quiescent germ cells are called at this stage. Some of these spermatogonia will simply produce more spermatogonia throughout the life of the man. However, other spermatogonia will be triggered to enter meiosis and produce four sperm each (Why four? Review meiosis in Figure 21 on p. 71). This process of spermatogenesis occurs within long tubules inside the testis; tubules that ultimately lead out to the sex ducts such that sperm are released to the outside during ejaculation. Within the testicular tubules the immature sperm are near the wall, and as meiosis is completed, the acrosome is formed, and the tail is produced, the sperm move steadily toward the middle of the tubule, where they are released so they can move out to the sex ducts. As mentioned above, sperm are thus produced continuously for the life of the man.

The process of producing eggs for fertilization by the sperm is a bit more involved but again is best thought of as being triggered by changes in the brain. Before the onset of puberty the female brain is not interested in sex hormone levels, but at puberty developmental changes occur that cause the brain to demand that the female sex hormone estrogen be present in the blood. The monitor of estrogen levels resides in a region of the brain that sits just above and behind the nose, called the hypothalamus. When puberty arrives and the hypothalamus first notices that estrogen levels are low, it produces a hormone, gonadotrophin-releasing hormone or GnRH, that travels a short distance down to the pituitary gland. In response to GnRH stimulation, the pituitary releases another hormone, follicle-stimulating hormone, or FSH, into the blood. As the name of this hormone indicates, its targets are the follicles of the ovary. When FSH reaches the ovary, it stimulates the follicle cells that surround the egg to both multiply and produce estrogen. This estrogen travels via the bloodstream to the hypothalamus, which responds to the rising estrogen levels by reducing its output of GnRH. However, the initial jolt of FSH incites developmental changes in the follicles that are not abruptly stopped just because FSH levels begin to fall. The follicle cells continue to multiply. Of course, many follicles are stimulated by the initial jolt of FSH, but their response is not uniform. Some follicles will grow a little bit faster than others. At some point in this developmental "race" one lead follicle reaches a critical stage of advancement before the others. This critical step is marked by an increase in the production of receptors for FSH. Thus the lead follicle harvests FSH from the blood much more efficiently than the other follicles, and it accordingly continues to grow and mature while the follicles that lose the race become deprived of FSH and recede. The dominant follicle continues to grow and produce estrogen, and during this process the egg resumes meiosis (recall that the eggs became developmentally arrested after the crossing over). The follicle cells also nurture the growth of the egg and produce the zona pellucida, which assists in sperm penetration, helps block fertilization by more than one sperm, provides a closed space that ensures appropriate interaction of cells in the cleaving embryo, and protects the embryo from infection by a variety of agents (see pp. 97–98 for a review of these functions). Estrogen levels thus continue to rise even though FSH is slowly falling. Figure 26 in Chapter 5 shows a developing follicle. Note that the follicle has a wall that clearly separates it from surrounding ovarian tissue. The egg, surrounded by follicle cells, sits in the middle of the follicle, and fluid begins to accumulate within the follicle as well.

At a critical point in this process estrogen levels rise above a threshold that triggers release of another pituitary hormone called luteinizing hormone, or LH. This surge of LH stimulates the follicle to rupture and release the egg into the fallopian tube, a response that requires about 36 hours to complete. When the egg is released from the ovary and into the fallopian tube for fertilization, it remains surrounded by a cloud of follicle cells. By this time meiosis has proceeded to the point where chromosomes are lined up for the second meiotic division. A first polar body is released, and there is no nuclear membrane surrounding the chromosomes. Once rupture of the follicle is induced by LH, cells within the ruptured follicle are altered such that they stop the production of estrogen and instead begin to produce the hormone progesterone. The progesterone-producing structure in the ovary is called the corpus luteum.

It is obvious that hormone stimulation must induce development and rupture of the follicles and stimulate growth of the egg as well as its progression through meiosis. These developmental changes in the egg are critically important for successful reproduction. However, the egg cannot be the only target for hormone stimulation. Not only must the egg progress to the point where sperm entry leads to a viable conceptus, but the lining of the uterus must also be conditioned to accept the early embryo and allow implantation and continued development. The most logical approach to developing the uterine lining in concert with egg development is to use hormones produced by the follicle for uterine stimulation. This is precisely what happens.

Before the onset of puberty the uterine lining is in a dormant state, but, once estrogen produced by the follicles enters the bloodstream, it reaches all tissues, including the uterus. Estrogen causes the cells in the uterus to proliferate and produce a thick, fluffy layer of cells. This is the first step in preparing the uterus to receive the blastocyst. After the LH surge and rupture of the follicle, the uterine lining becomes exposed to progesterone from the corpus luteum of the ovary, and this exposure completes the conditioning process. At this point in the reproductive cycle either of two things can happen: A blastocyst may implant into the uterus for continued development, or the reproductive cycle may not be associated with successful implantation. Let's consider the latter of these situations first. After release of the egg, the ovary produces progesterone, which completes maturation of the uterine lining and maintains it in a state suitable for implantation. But the ovary does not want to continue maintaining the uterine lining indefinitely. Rather, it wants to initiate a new cycle of ovulation in order to "try again" for conception in the event of fertilization failure. Accordingly, the ovary continues producing progesterone for about 2 weeks, and if implantation does not take place the corpus luteum recedes and progesterone production stops. Withdrawal of progesterone from the endometrium causes the lining to slough off, a process we recognize as the menstrual period. Once menstruation occurs, estrogen levels and progesterone levels are low. The hypothalamus responds to this situation by renewing its demand for estrogen, and a new FSH surge occurs. A new ovulatory cycle thus begins.

The second situation occurs when fertilization does take place. The task of the developing embryo is to prevent the ovary from giving up the production of progesterone, a process that leads to shedding of the uterine lining, or endometrium, during menstruation. Menses would be fatal to the embryo because it would lead to its expulsion from the body. The embryo ensures continued progesterone production by producing a hormone that functions in essentially the same way as LH. LH from the pituitary stimulates progesterone production from the ovary. The embryo's progesterone inducer, which works in essentially the same way as LH, is called human chorionic gonadotrophin, or HCG. HCG production begins even before implantation, but shortly after implantation it reaches levels that "rescue" the corpus luteum from demise and induces it to continue progesterone production. Because HCG is much like LH, it can also induce follicle rupture if administered at the appropriate time to a human or an animal. The old tests for pregnancy involved injection of urine from a woman into a female rabbit. In the rabbit, the ovary sits in a constant state of readiness for LH stimulation. Accordingly, when the urine from a pregnant woman is injected, the HCG in the urine induces the rabbit to ovulate. These tests have since been replaced by far more sensitive chemical assays for HCG, but they illustrate the similarities between HCG and LH.

In Vitro Fertilization and Related Assisted Reproductive Technologies

In vitro fertilization, or IVF, was originally developed for treatment of female infertility resulting from conditions such as blocked fallopian tubes. However, these technologies assume a central role in efforts to engineer the human germ line. These procedures make fertilized eggs available for manipulation and allow for production of far more eggs than the single egg typically produced from the dominant follicle in a normal reproductive cycle. The general strategy for in vitro fertilization is to obtain eggs from the ovary just before ovulation, expose them to sperm in the test tube, and transfer fertilized eggs into the uterine cavity from below (that is, through the cervix) for continued development. It is quite apparent how this approach would circumvent blockage of the fallopian tubes. However, as these technologies have developed their applications have diversified remarkably.

A logical approach to in vitro fertilization is to measure hormones in the woman and monitor rising estrogen levels that indicate the presence of a developing follicle. Ultrasound can then be used to identify the dominant follicle, which is readily visible as fluid accumulated within it (see Figure 26 in Chapter 5). If this process is carefully watched, it is possible to aspirate the egg from the dominant follicle within the 36-hour interval between LH release and follicle rupture. The egg can then be fertilized in vitro and returned to the uterus through the cervix. Although this strategy is logical, it works poorly because only a single egg can be retrieved and because the timing of egg retrieval is so critical.

A major step forward in IVF technology was made when hormones were used that allowed maturation of many follicles rather than only one. In a normal cycle, the dominant follicle emerges and deprives the other follicles of the FSH essential for continued maturation. When an IVF is performed, the woman is injected with large quantities of FSH. This allows follicles to develop that normally would have fallen by the wayside. The result is emergence of many follicles with mature oocytes that are capable of being fertilized. The development of multiple large follicles results in higher levels of estrogen production than would occur in a normal cycle and the appearance of many fluid-filled maturing follicles on ultrasound examination.

The use of FSH to sustain development of multiple follicles is an important feature of IVF because it allows for retrieval of multiple eggs. Another important step in IVF management is control of the timing of ovulation. This is done by administering LH to the woman before the time that the brain would release it. Typically HCG is used, because it has the same actions as LH. If HCG can be given before the woman's own endogenous surge of LH takes place, the IVF practitioner can be certain of the timing of follicle rupture and can aspirate eggs from the ovary a few hours before spontaneous rupture. This approach optimizes oocyte maturation and prevents loss of eggs that would occur if ovulation took place. After eggs are surgically retrieved from the ovary, they are allowed to mature for a few hours to compensate for their slightly premature removal from the follicles. They can then be exposed to sperm in vitro, examined for fertilization and cleavage development, and returned to the uterus via the cervix for establishment of pregnancy. One must be careful at this point not to transfer large numbers of embryos back to the uterus. To do so would incur the risk of a multiple pregnancy, a potential disaster for both mother and fetuses. Presently, it is unusual for more than three embryos to be transferred in a single IVF cycle. Additional embryos are typically frozen and used for transfer at a later time in the event pregnancy does not occur with the first transfer or for conception of more children in cases where the couple wishes to conceive children additional to those born from the initial retrieval cycle. When three embryos are transferred rates for establishment of successful pregnancy vary between laboratories, but the average success rate is about 20%.

When frozen embryos are transferred, it is of course not necessary for the woman to be stimulated with hormones. All that is required is careful timing of maturation of the endometrium, such that embryos are thawed and transferred during the window of implantation. It should be clear from this scenario that the woman who receives the embryos need not ovulate herself. It is important only that her endometrium be properly prepared to receive cleaving embryos. Because any properly prepared uterus can sustain implantation of the transferred embryo and allow pregnancy to continue, surrogate mothers can be enlisted to lend a uterus for embryo donation in the event that the woman from which the eggs are obtained has a condition that does not allow her own uterus to function properly. The surrogate mother who carries the embryos must of course have a uterus, but she need not have functional ovaries: The estrogen and progesterone needed by the endometrium can be supplied from the outside. This capability has allowed birth of children from postmenopausal surrogate mothers. From this discussion it should be apparent how IVF can be used to address a variety of reproductive disorders in women. Even the absence of a uterus does not preclude successful reproduction.

What can assisted reproductive technologies accomplish for male infertility? First, it should be appreciated that IVF allows for exposure of eggs to many more sperm than is possible in the fallopian tube. If a man has relatively a small proportion of normal sperm, the use of high numbers of sperm in IVF can compensate for this problem. In general, however, interaction of sperm and eggs in the test tube is less efficient than in the fallopian tube, and IVF alone cannot compensate for severe problems of male infertility.

If we recall the process of sperm penetration, we realize that although the zona pellucida induces the acrosome reaction it also constitutes a potential barrier to sperm penetration (see p. 97). Suppose we circumvent the zona barrier. There are several ways of doing this. First we could simply open a hole in the zona and allow sperm to swim unobstructed to the oocyte surface. Of course, only acrosome-reacted sperm will penetrate the egg, and those sperm that swim through a hole in the zona will not have had the benefit of zona binding that induces the acrosome reaction (see pp. 97–98). However, as it turns out, many sperm undergo the acrosome reaction spontaneously, and when a hole is opened in the zona these spontaneously acrosome-reacted sperm have a much easier time reaching the egg surface. As a consequence, opening the zona can alleviate problems of male infertility that result from low numbers of normal sperm. Another approach can entail loading of sperm into a microneedle, insertion of the microneedle under the zona, and squirting several sperm under the zona. This manipulation works on the same principle as zona opening. Both zona opening and subzonal sperm insertion naturally run a risk that multiple sperm will penetrate the egg, because the function of the zona in blocking polyspermic fertilization is negated when the zona is bypassed. Ideally, a procedure for treating male infertility would result in efficient fertilization by only one sperm.

This objective was reached when it was discovered in the early 1990s by a research group in Belgium that single sperm could be injected directly into the egg. When this is done, the egg is induced to complete meiosis and the sperm decondenses to form a pronucleus. It is truly remarkable that the multiple interactions between sperm and egg in the normal fertilization process can be completely circumvented and normal fertilization achieved. It appears that even when sperm with their acrosomes intact are injected by this process, termed intracytoplasmic sperm injection, or ICSI, fertilization and normal development can still take place. Amazingly, sperm do not even have to mature to be successfully injected. One can retrieve maturing sperm directly from the testicle, perform ICSI, and obtain babies. In mice it has even been possible to freeze-dry sperm, store them on the shelf like a chemical reagent, use them later for ICSI, and obtain normal live young. There is little reason to believe the procedures would not work in humans as well.

Before we move on we can now answer the question posed at the head of this chapter. How can a man who has had a vasectomy have children with a woman who has had a hysterectomy? Simple: We induce ovulation in the woman and retrieve oocytes for IVF. We aspirate sperm from the man's testicle and fertilize the woman's eggs by ICSI. We then transfer the embryos into the uterus of a surrogate mother whose uterus has been synchronously prepared for implantation of the embryos. The surrogate mother receives three transferred embryos, two of which implant and become a boy and the girl, who are then returned to the biological mother. Before we move on to other uses of IVF and embryo manipulation, we should emphasize that we are not discussing the ethical implications of such a strategy for conception, we are only pointing out its feasibility from the scientific perspective. Ethical issues will be visited later.

Genetic Diagnosis on Embryos

The availability of multiple preimplantation embryos as part of the IVF procedure, the amenability of embryos to a variety of manipulations, and advances in molecular biology have made possible the diagnosis of genetic disease in the preimplantation embryo. As I showed above, embryos can have as many as 3/4 of their cells removed without compromising their ability to develop into live young (see pp. 104–105). We also know that PCR has made possible the amplification of very small quantities of DNA to produce amounts suitable for direct sequence analysis. These realities should make it clear how genetic disease can be diagnosed even before embryos implant into the uterine wall. Genetic diagnosis at such an early stage of development can allow embryos with genetic disease to be excluded from transfer, thus avoiding the decision to abort established pregnancies. The approach is straightforward: Simply open the zona, remove a cell from the cleaving embryo, perform PCR on the DNA of that cell for regions of genome that are suspected to carry disease mutations, analyze the sequence of the amplified DNA, and determine which embryos have two, one, or no copies of a disease-causing mutation. This analysis can be completed in such a short time that the answer can be obtained before it is necessary to transfer or freeze the embryo. With this result in hand, embryos known to have a complement of genes that will not lead to disease can be selectively transferred. Again, no attempt is made here to discuss the ethics of this process. Suffice it to say for now that this preimplantation genetic diagnosis, or PGD, is technically feasible and has been used successfully for avoidance of genetic disease.

7

METHODS AND STRATEGIES FOR GENE TRANSFER AND ENGINEERING OF THE GERM LINE

"Genetic transformation of mouse embryos by microinjection of purified DNA"
—Title of a 1980 research publication by J. W. Gordon, G. A. Scangos, D. J. Plotkin, J. A. Barbosa, and F. H. Ruddle

Information on genetics development and reproductive biology has been presented above for one purpose—to make the various approaches to genetic manipulation readily understandable and to provide an appreciation of the relative advantages and limitations of each approach. There are a variety of strategies for engineering the genome to produce offspring with specific desired characteristics. Some of these processes are straightforward, involving simple selection for inheritance of patterns of alleles that confer a desired appearance, or "phenotype." Other methods involve insertion of new genetic material, replacement of existing genes with new alleles of the same gene, or even removal of genes. These latter processes require a knowledge of methods of gene transfer, which is easily attained, and an understanding of reproductive biology, developmental biology, and genetics, which has already been provided. Let's first consider the direct approach of genetic selection.

The Science and Ethics of Engineering the Human Germ Line: Mendel's Maze, by Jon W. Gordon
ISBN 0-471-20647-4 Copyright © 2003 John Wiley & Sons, Inc.

Selective Breeding

The easiest way to select for desired patterns of allelic inheritance is to simply optimize the reproductive potential of organisms that display desired characteristics while limiting reproduction of organisms that do not have the desired phenotype. If one is breeding cattle for high milk production, it is a simple matter to measure milk production in females and allow those that produce the most milk to have calves. If this process is continued over many generations, it is possible to produce cows that yield many times more milk than their ancestors. This process of genetic selection assumes that the desired traits have a genetic basis, but it makes no assumptions about how many genes might be involved in generating the desired trait. The method is very reliable, but the results require a long wait. Genetic selection in humans would not be a very practical approach to genetic engineering. Breeding partners in humans are not chosen on the basis of desired traits but a variety of social and personal factors that rarely have much to do with the genetic endowment of the individuals involved. In addition, those who want to engineer their offspring to have desired traits do not want to wait for 20 generations to see the results.

A new opportunity for selecting desirable combinations of genes in human offspring has been created by the human genome project. This effort entails the sequencing of the entire three billion base human genome, followed by an analysis of the sequencing data. The complete human gene sequence was deduced in 2001, and it is already being used to discern the genetic basis, when one exists, of disease traits. Here is how the analysis works.

When one has in hand the entire human genome sequence, one can identify DNA sequences that are specific for each of the 23 chromosomes. In fact, possession of the entire human sequence allows assignment of specific DNA fragments to subregions of any chromosome. In many cases these DNA fragments are not parts of genes but rather are present in spacer DNA that is situated between genes. Nonetheless, these fragments are assigned to chromosomes, and slight variations in the sequences of these fragments can frequently allow them to be traced specifically to a chromosomal region inherited either from the mother or the father.

With this information in hand, let us now consider a common disease such as Parkinson disease, a neurodegenerative condition that causes problems of muscle control. Parkinson disease is quite common, and there is little evidence from identical twin studies that any one specific gene is responsible for it. However, if a large number of patients with Parkinson disease are examined for association of the condition with a specific region of a chromosome (an association that can be made by screening DNA from individuals with the disease for DNA marker fragments from every chromosome), it is possible to identify markers present in chromosomal regions that are inherited by patients with Parkinson disease with a higher frequency than would be predicted from random assortment. The DNA marker fragments used for this analysis need not be part of a gene that predisposes to the disease; they need only be closely linked to such a gene or genes. Remember from the discussion of crossing over that a fragment a million bases distant from a gene on a chromosome will only be separated from that gene by crossing over 1% of the time (see discussion of crossing over on p. 79). Therefore, although this analysis does not pinpoint genes that are involved in risk for Parkinson disease, it does determine that various regions of specific chromosomes have a gene or genes that contribute either positively or negatively to this risk. It is important to understand here that none of the genetic elements associated with Parkinson disease by this analysis are invariantly associated with development of the disease; rather, they are found with increased frequency in affected individuals. They are not found in all affected individuals, as is the case with disease inherited strictly as single-gene traits such as Huntington disease, Tay–Sachs disease, or sickle cell disease.

As analyses such as these move forward, it will undoubtedly be possible to attribute relative risk for a variety of disease states to regions of chromosomes and, possibly, eventually to specific genes within those chromosomal regions. Assuming that this knowledge will be accumulated, it is easy to see how prospective parents could be screened for possession of combinations of allelic variants that would be associated with increased or decreased risk of various diseases in their children. Moreover, positive selection might be made for combinations of alleles that would favor specific physical characteristics, and perhaps even emotional or intellectual characteristics, in children (although the latter traits, as we shall see later, may be very difficult to define genetically). It is this type of selection that is portrayed in the science fiction film *Gattaca* (1997). In the scenario presented in this film, highly intelligent, physically strong, and attractive persons are born through a program of genetic selection.

Although it is true that with the relevant knowledge in hand this kind of genetic selection could be made within a single generation, the strategy also has severe limitations. First, optimization of such a strategy would require a comprehensive knowledge of chromosome regions and associated markers for all of the traits of interest. In the absence of such knowledge several generations of crossing would be required to assemble in one child the combinations of alleles desired. Another significant limitation is that genes are inherited in blocks. Recall that two DNA fragments one million bases apart on the chromosome are separated only 1 percent of the time by chromosomal crossing over. Although this fact can be exploited to associate DNA markers to chromosomal regions that influence specific traits, it is also unavoidable that these markers will be linked to allelic variants of other genes that might be disadvantageous with regard to other traits. Just as the markers used are difficult to separate from genes of interest, they are difficult to separate from other nearby genes that may be of a disadvantage to the child. To address this problem it would be necessary to screen perhaps millions of genomes to identify markers that are associated only with desirable alleles of nearby genes.

Another problem is that there simply aren't as many genes in the human genome as was originally thought. Although enough DNA exists to encode a million genes, results from the human genome project thus far indicate that there in fact exist about 40,000 genes. With so "few" genes encoding the myriad structures and cellular functions of an adult human being, it becomes almost a certainty that individual genes can have effects on many different traits. Thus alleles favorable for one phenotype may actually be unfavorable with regard to another phenotype. If this likely possibility ultimately proves to be the case, it might be impossible to successfully select for an "ideal" pattern of phenotypic traits.

A final important limitation to this selection approach is that chromosomal markers for complex traits may predict an increased or decreased probability that a specific phenotype will appear in the offspring, but this probability is by no means a certainty. A desirable pattern of alleles for a reduced likelihood of Parkinson disease might be selected in a child by genetic diagnostic procedures performed on parents and embryos or fetuses, but inheritance of the desired pattern of alleles would not guarantee that the child will not develop Parkinson disease. When a strict genotype-phenotype relationship cannot be established for a gene or chromosomal region, as it can for traits such as Huntington disease, then selection becomes a game of probability. And, again, optimization of this approach would necessitate a method of choosing genetic parents that deviated from the current custom of doing so on the basis of emotional ties between father and mother. Although the limitations of genetic selection cannot be minimized, results thus far generated from the human genome project do indicate that selection might be used for avoidance of a dramatically increased risk of one or perhaps a few disease traits that are influenced by genotype. In a family with a strong history of early-onset heart disease, the ability to link that disease propensity to a region or regions of chromosomes, and to screen embryos by PGD or fetuses by amniocentesis for the absence of these regions, is rapidly approaching on the medical horizon.

Basic Gene Transfer Methodology

Given the limitations of genetic selection as an approach to modifying the human germ line, it is not surprising that more powerful and direct approaches have been sought. These approaches are all based on our ability to insert foreign DNA sequences directly into the genomes of embryos. Gene transfer can be used to add new traits, modify existing traits, or even negate the function of genes. To understand how such powerful technologies can be extended to the germ line, we must first acquaint ourselves with DNA transfer methodology. This examination is best undertaken by first looking at transfer of genes into tissue culture cells.

It is surprisingly easy to introduce new genetic material into cells. Of the many methods available for doing this, one of the most popular is electroporation. This is a simple procedure in which cells are placed in suspension, DNA is added, and the mixture is exposed to a pulse of electricity. The electrical pulse creates holes in the cells' membranes, which allow DNA to enter. DNA then makes its way to the nucleus, where it integrates into one of the chromosomes. Electroporation is not very efficient. Although rates of gene transfer success vary between different cell types and protocols, a reasonable average estimate is that 1 in 1000 cells subjected to DNA transfer will actually take up and express the donor gene. Given these low success rates, a method of identifying the cells that take up the genes is clearly needed. A very popular method for identifying genetically "transduced" cells is including in the donor DNA preparation a gene that will allow cells that express it to survive exposure to a poison. Nowadays the most common system for such selection involves the drug neomycin, which is poisonous to mammalian cells. Some bacteria have a gene that encodes resistance to neomycin toxicity, a gene we shall abbreviate as Neo^R. When gene transfer is performed, the gene of interest is mixed with or directly linked to the Neo^R gene. After electroporation is performed, the cells are allowed to recover and are given a couple of days to turn on the Neo^R gene, after which neomycin is added to the culture medium. Neomycin then kills all cells except those expressing the Neo^R gene, and thus colonies of cells that emerge during this subsequent culture period all carry the new gene. Note that the Neo^R gene need not be physically linked to the gene of interest; it can simply be mixed with it before electroporation. For reasons still not entirely clear, separate DNA fragments almost always become linked during the gene transfer process. It should also be appreciated that the widely dispersed colonies of neomycin-resistant cells that appear after culture in the drug are each a group of descendants of a single successfully transduced cell. Therefore, the site of gene insertion within the cells of each colony will be the same, the structure of the foreign DNA fragment will be the same, and the level of expression of donor genes will be the same. You might wonder where in the chromosomes foreign DNA integrates after gene transfer. In general, the site of insertion of the foreign DNA appears random, except under very rare circumstances as discussed below.

Nonrandom, or "targeted," integration of foreign DNA can occur in tissue culture cells if conditions are right. If we transfer a mouse globin gene into a mouse cultured cell, there will be two genes in the host cell that carry an identical, or nearly identical, sequence to the globin gene introduced from the outside (we say "nearly identical" because sometimes there will be small variations in the sequence, as is the case for human sickle globin and its nonmutated counterpart). When a donor fragment with a sequence highly similar to a gene segment in one of the chromosomes is introduced into the cell, it will occasionally line up with the chromosomal sequence, in much the same way that chromatids line up for crossing over on the basis of sequence homology (see pp. 76–78). Once the donor and host sequences become aligned, a "double crossover" event can take place, such that an internal segment of the donor fragment is "swapped" for its counterpart in the chromosome. This event is depicted in Figure 33. Of the 1 in 1000 cells or so that take up foreign DNA after electroporation, on average only 1 in 1000 of these will exhibit this kind of targeted integration, and of course, such targeting can only occur if the added DNA has a sequence highly similar to one found in the host chromosomes. Although rare, targeted integration events are highly important because they are not random and because the behavior of the donor DNA molecule is likely to be quite similar to its chromosomal counterpart after gene transfer. As we shall see, these features of targeted integration make it a highly preferable method for introducing new genetic material into the human germ line.

Given the potential importance of targeted integration events, how can we expect to find them when they occur in only about one in every million cells exposed to the DNA? Although this task sounds daunting, it is in fact not so difficult. First, it should be appreciated that even though only 1 in 1,000,000

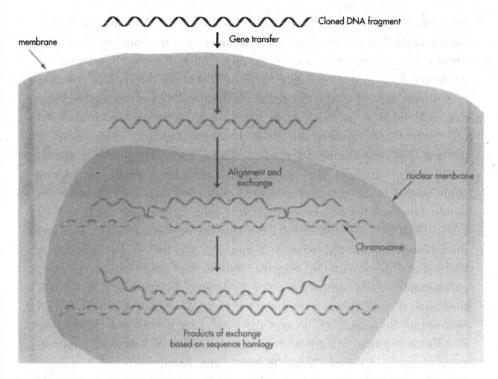

Figure 33 Targeted DNA integration after gene transfer into a cultured cell. Note that a double break and exchange leads to replacement of a recipient segment of the chromosome by the donor fragment.

cells undergoes this special form of gene insertion, it is quite easy to expose 100,000,000 cells to foreign DNA in a single electroporation experiment. That means that if we could only find them, there would be 100 cells with targeted integration produced from one gene transfer effort. Therefore, although rare, these events are common enough to find routinely. There are several workable methods for identifying cells with targeted integration. An example of one method, given below, employs the specificity of PCR (see pp. 51–52, Chapter 3) and the use of the selectable Neo^R gene. It works as follows.

We will first engineer our donor DNA fragment with gene splicing technology in such a way that the Neo^R gene is near one end of the gene. We then choose a gene for transfer such that we know the DNA sequence not only of the electroporated fragment, but also of the several hundred bases of DNA in the chromosome that immediately flank this fragment when it is sitting in its native site in the chromosome. When targeted integration occurs, the crossover event can take place in such a way as to involve the DNA of the donor fragment that lies between the Neo^R gene and the end of the fragment. The exact site of the other crossover event near the other end of the donor fragment is not important. If you examine Figure 33 you will see how this kind of crossover even places the Neo^R gene very close to the flanking DNA of the targeted chromosome; and as we said before, we know the sequence of the chromosomal DNA immediately flanking the site of the donor fragment "swap." Now we perform PCR with primers designed as follows: One of the primers is specific for 20 bases or so of the Neo^R gene itself, and the other is specific for flanking chromosomal DNA that, after a targeted integration event, will be just a few hundred bases away from the inserted Neo^R gene. Under these circumstances, PCR will amplify the segment of DNA that sits between the Neo^R gene and the flanking chromosomal sequence (Figure 34, lower left). It should be easy to realize that if the donor DNA fragment with its Neo^R gene lands on any other chromosome, or at a nonspecific site of the targeted chromosome that is tens of thousands of bases distant from the site of targeted integration, the PCR reaction cannot be completed, because the Neo^R gene primer and the flanking chromosomal primer will not be close enough to allow PCR to work. This outcome is shown in Figure 34 on the lower right. Recall that PCR only works when the primers are at most a few thousand bases from each other. It should thus be clear that in the 999 of 1000 genetically transduced cell clones with nontargeted integration, a subsequent PCR reaction with such primers would yield no product at all. However, after targeted integration, a PCR product of predictable size and sequence will appear. One can accordingly sample genetically transduced clones, pool them in groups of one hundred, and perform PCR. If a product appears, then at least one of the clones in that pool has targeted integration. Such pools can thus be quickly subdivided and rescreened so that the clones with targeted integration are isolated. Remember that the cell colonies were only sampled for DNA isolation and PCR analysis. Other cells from each colony can either be cultured or frozen during the analysis period. When protocols such as these are followed, the cells with targeted integration invariably have a sequence consistent with a perfect swapping event as shown in Figures 33 and 34.

Gene Transfer into the Germ Line in Animals

It should be readily apparent that gene transfer protocols such as electroporation and neomycin selection are not readily applicable to insertion of genes

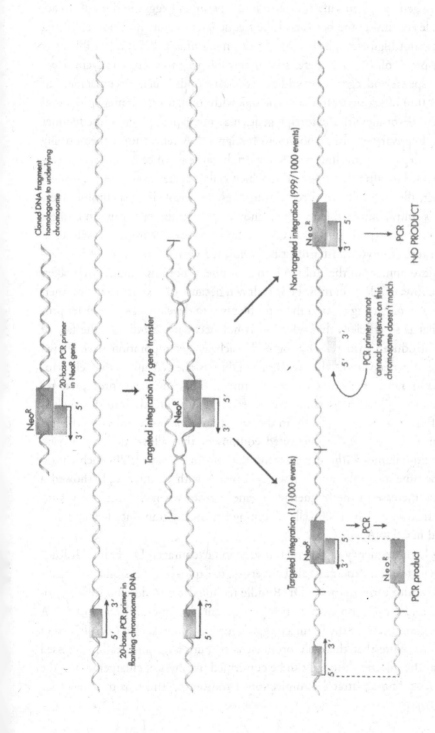

Figure 34 Detection of targeted DNA integration by PCR. One PCR primer is in the targeted chromosome outside the area of DNA interaction. Only if the Neo^R gene lands near this flanking site can PCR be completed (compare lower left with lower right).

into sperm, eggs, or early embryos, the logical gene transfer targets when one wishes to genetically modify the germ line. Sperm and eggs are difficult to access while still inside the body, and the frequencies of gene insertion discussed above are not high enough to allow development of a technically feasible gene transfer protocol. Because there is no reasonable way to exert neomycin selection on sperm and eggs, it would be necessary under such circumstances to screen 1,000 offspring to find a single one with foreign DNA integration, and those few offspring with the foreign genes may not express them in the manner desired. Exposure of early embryos to foreign DNA represents a theoretically viable strategy for germ line gene insertion, because germ cells arise in the embryo many days after the time during which gene transfer would be performed. However, the frequency of success using these protocols is again prohibitively low: It is simply impossible to obtain thousands of embryos for use in a single electroporation experiment, and very few days are available for neomycin selection because embryo transfer must be performed within a few days of fertilization if development of the embryo is to continue. Once implantation has taken place neomycin selection must be withdrawn because of toxicity to the mother.

One way of getting around these problems is to develop a gene transfer protocol that is so efficient that selection is not necessary. Studies of the fate of DNA introduced into cells by methods such as electroporation have shown that the vast majority of the transferred DNA is chewed up in the cell cytoplasm and never reaches the nucleus intact. This discovery prompted researchers to explore the possibility of injecting DNA directly into the nucleus in an effort to avoid degradation in the cytoplasm. Of course, injecting the nucleus of a cell requires sophisticated equipment that allows delivery of very small fluid volumes without destroying the cell. In the late 1970s such equipment became available, and a few experiments with cultured cells showed a dramatic increase in the frequency of gene transfer when the nucleus was injected: Instead of 1 in 1000 cells becoming genetically transduced, 1 in 5 cells retained new DNA.

This high frequency of genetic transduction encouraged Dr. Frank H. Ruddle, a gene transfer expert at Yale University, to explore the possibility of injecting DNA into early embryos. Dr. Ruddle recruited me to develop such an approach, and I came up with a strategy that entailed microinjection of DNA into the pronucleus of the fertilized egg. Figure 35 shows such a microinjection procedure. Note that the male pronucleus is quite large and readily accessed and that the injection process can be confirmed by physical enlargement of the pronucleus. Shortly after microinjection, I transferred embryos to females for

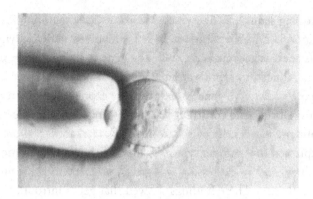

Figure 35   Injection of genes into the fertilized egg. A suction pipette holds the embryo in place (left) while a microneedle (right) injects DNA into the male pronucleus.

continued development. Immediate transfer was performed to minimize potential harmful effects of in vitro culture. We produced about 70 newborn animals in our initial experiments and demonstrated that 3 of the offspring in fact had retained the foreign DNA in what appeared to be all of their cells. This first successful insertion of recombinant DNA into the embryo was reported in a publication whose title is shown at the head of this chapter. A follow-up study published soon after involved insertion of the first human gene into an animal and demonstrated both integration of the new genetic material into a mouse chromosome and transmission of that material to subsequent generations of offspring through breeding. Immediately after our first publication many laboratories turned their attention to this new gene transfer approach, and similar findings appeared from several research groups at about the same time. We subsequently called these animals "transgenic" mice, a term that is now used to describe virtually all organisms that receive foreign genes.

The usefulness of transgenic mice as research tools was critically dependent on the behavior of the new genetic material. For this methodology to realize its full potential, it was important that the new introduced genes be expressed efficiently. About a year after our first transgenic mouse was reported to the scientific community, two other scientists, Ralph Brinster of the University of Pennsylvania and Richard Palmiter of the University of Washington at Seattle, produced a transgenic mouse that addressed this concern while at the same time answering one of the most profound questions concerning mechanisms of gene regulation. They used recombinant DNA technology to link the promoter region of a mouse gene, metallothionein-1, abbreviated MT-1, to a herpes virus gene coding sequence. The viral gene was called thymidine kinase, and it is involved in viral DNA replication. MT-1 is involved in detoxifying ingested heavy metals such as cadmium or zinc. The gene is active mainly in liver and kidney and is induced to express at high levels if an animal is exposed to heavy metals. Of course, the viral gene is not normally expressed in any mouse cell.

When this hybrid gene construct was used to make transgenic mice, it was of course present in all cells of the animals. Remarkably, however, the viral gene was expressed in a manner typical of the MT-1 gene: It was most active in liver and kidney, and its expression was greatly augmented if the animals were exposed to cadmium. These findings showed that genes introduced by pronuclear microinjection could be very efficiently expressed in adult animals and further showed that DNA regions situated very close to the coding sequences were responsible for determining the specific pattern of gene expression. These findings were quickly exploited by Palmiter and Brinster to produce a variety of animals with MT-1 regulators linked to various other genes. Mice with MT-1/growth hormone genes grew much faster and to twice the size of normal mice, and mice with genetic growth hormone deficiency were corrected in their growth by the transgene. In all of these experiments the new transgene was produced from tissues typical of MT-1 expression rather than from the tissues that normally elaborated these gene products. Thus, although growth hormone is normally produced from the pituitary gland, growth hormone produced by MT-1/growth hormone transgenes was elaborated from the liver.

These experiments not only established the potential for pronuclear microinjection to obtain efficient expression of genes introduced into embryos, it solved one of the major puzzles in developmental biology. Before transgenic technology, it was unclear why genes present in every cell, like globin, were active in only a single cell type, the red blood cell. The experiments in which the region of the MT-1 gene promoter, when grafted to other genes, could direct expression of those other genes to express in a pattern typical of MT-1 showed that small regions of DNA close to genes were responsible for activating those genes in the appropriate tissues. In the case of MT-1 the key DNA sequences are where the promoter lies, on the immediate 5' side of the gene. But further transgenic research showed that regions that confer tissue specificity of expression on genes can be within the noncoding intron regions that are removed during processing of the mRNA (see p. 39 for a review of processing) or they may be several thousand bases away on either side of a gene. These small regions of DNA are called "enhancers," because they enhance expression in specific tissues. Research characterizing enhancers of a number of genes has also shown that a gene expressed in only a single cell, like globin in the red blood cell, may be associated with several enhancers located on both sides of the gene as well as within it. Powerful enhancers for globin gene expression in red blood cells may be as far as 50,000 bases away from the gene. The enhancer regions were identified by showing that in red blood cells, where globin is expressed, the DNA in the globin region is more exposed and not sequestered by complexes of proteins that ordinarily maintain DNA is a quiescent state (remember that shutting down genes and keeping them off is a major imperative for the developing organism). It appears from transgenic and other experiments that globin genes are complexed with protein and silent until the adult red blood cells begin to form late in fetal development. Then the globin enhancers expose the DNA and allow expression to begin. The fact that discrete enhancer regions confer tissue specificity of gene expression has been exploited to achieve novel patterns of gene expression in transgenic animals. One can graft globin enhancers to a gene normally silent in red blood cells and obtain expression of that gene in red blood cells. An interesting characteristic of enhancers that makes them very convenient for use in genetic engineering is that they exert their enhancing effects if you can just get them close to a gene. It doesn't matter if an enhancer normally located "upstream" of a gene (on the promoter side) is moved to a site "downstream" (past the transcription termination signal) of another or even of its own gene; it still works just as well. Moreover, you can flip enhancers over, such that the side closest to the gene becomes switched for the side normally located further away, and the enhancer still works fine. In fact, part of the definition of enhancers is that they function independently of their position or orientation relative to the coding sequence of the enhanced gene. What a marvelous convenience for designing genes to express in novel patterns! It's so much easier to do gene splicing when the rules for producing a functionally expressing hybrid gene are so relaxed.

One might expect anomalous patterns of "transgene" expression produced by exogenous genes in transgenic mice to cause serious developmental problems for the transgenic animal. This is often true, but it is also surprising how well many transgenic animals, usually mice, tolerate high levels of expression of genes in tissues where these genes are normally inactive. I say "usually" mice because it appears that pronuclear microinjection can be extended to any mammal in which pronuclear-stage embryos can be retrieved or created by in vitro fertilization, microinjected, and returned to females for continued development to term. It has thus far been possible to extend pronuclear microinjection to sheep, cattle, goats, rabbits, and rats. Although the procedure is somewhat less efficient for many of these species than it is for mice, it can be used introduce genes into the germ line in all of them.

It is not possible here to review the many biological problems solved by the advent of transgenic mouse technology, but a couple of examples will serve to illustrate the enormous power of this new tool. Consider the fact that in many cancers such as breast cancer genes that promote cell growth or division are expressed at abnormally high levels. This observation could indicate that the primary defect in these malignant cells is that they overexpress growth-promoting genes and the abnormal expression of these genes induces uncontrolled tumor growth. However, it could also be that the cancers develop from some other mechanism and that the activation of the growth promoting genes is a secondary event that may help sustain tumor growth but does not cause the tumor to arise. With transgenic mice it is possible to introduce these growth-promoting genes linked to enhancers that cause marked overexpression in breast tissue. The animals are then born with this gene expressed in very large quantities in their normal breast tissue and observed for the development of breast cancer. The result of this experiment is that the transgenic mice develop multiple breast tumors, a result that strongly implicates aberrant expression of the growth-promoting genes as the cause of the cancer. This finding allows researchers to focus attention on events that cause these genes to become expressed at abnormally high levels.

As another example, consider again the bleeding disorder hemophilia, which was mentioned previously. This is a group of diseases resulting from mutations in genes that produce proteins needed to clot the blood. Before transgenic technology, hemophiliacs were treated by pooling precious clotting factors isolated from thousands of blood donors and administering the factor during acute bleeding episodes. The problem with this form of treatment was that human blood donors often had serious infectious diseases such as AIDS or hepatitis, and hemophiliacs were placed in the terrible situation of either suffering crippling and life-threatening bleeding episodes or contracting potentially fatal infectious diseases. When transgenic technology came along, it was possible to link enhancers from genes that encoded milk proteins, and that expressed at very high levels in breast glands, to human clotting factor genes. These hybrid constructs could then be used to produce transgenic goats or sheep, and when female transgenic livestock matured it was possible to go to the barn and milk these animals to obtain clotting factors completely free of the risk of infectious disease contamination. Although better treatments for hemophilia have since been developed, the enormous potential of transgenic technology to engineer animals as "bioreactors" for production of valuable, complex proteins was demonstrated. As already mentioned, this methodology can also be used to produce animals with more rapid and extensive growth by overexpression of growth hormone transgenes. The potential of this strategy for engineering live-stock with more rapid growth will be discussed later. Since we developed the first transgenic mouse in 1980, more than 10,000 research studies utilizing transgenic technology have been published.

One might think that with the advent of pronuclear injection the problem of devising an approach to human germ line genetic engineering was solved. After all, about 1/6th of all embryos that develop after pronuclear injection have an integrated transgene and the genes are very highly expressed. Why not cure sickle cell anemia by injecting the normal globin gene into embryos that have two copies of the sickle gene only, obtaining high expression, and produc-ing a child that has enough normal globin to function as a carrier of the sickle gene rather than an individual with pure sickle cell anemia? What could be eas-ier? It is easy to obtain human embryos by IVF, and the pronucleus in the hu-man fertilized egg is very readily visualized.

As we shall see in our later discussion of the differences between procedures performed on experimental animals and those performed on humans, many is-sues come into play when humans are involved that simply do not apply to an-imals. Just to mention one example here, a success rate of 1/6 is fine for mice but not very good when extended to humans, where 5/6th of the patients will suffer treatment failure. However, without delving deeply into the problems re-lated to the clinical application of such procedures right now, there are many other more fundamental problems with using pronuclear microinjection on human embryos.

When genes injected into the pronucleus are examined in the resultant transgenic animal, they often exhibit many structural and functional features that, if present in a human, would pose great risk for more harm than good. It is first important to remember that genes injected into the pronucleus are inte-grated randomly into the host chromosome rather than in a targeted fashion. Extensive efforts to identify targeted integration events after pronuclear mi-croinjection have uncovered evidence suggestive of targerted integration in fewer than a handful of the tens of thousands of animals produced. When inte-gration occurs randomly, it has the potential to profoundly disturb the func-tion of other genes resident within the chromosomes of the recipient organism. Consider, for example, a situation in which the globin gene inserted itself into the middle of a clotting factor gene. The new transgenic individual would probably express the globin transgene, but the reading frame of his/her clotting factor gene would be terribly disrupted and the clotting factor gene would be effectively destroyed. This kind of event is called an "insertional mutation," and in animals is detected about 5% of the time. I say "detected" because there may be insertional events that disturb host gene expression but do so in ways too subtle for detection in an animal. Whereas insertion directly into the coding sequence of another gene yields a readily detectable abnormality, other insertions, which might destroy enhancers or separate them from their host coding sequence, could cause alterations in gene expression that would not harm an animal sufficiently to allow detection but which, in humans, could yield unacceptable problems such as developmental disabilities. Random integration is therefore a serious impediment to the use of pronuclear microinjection in humans.

Another feature of microinjected genes is that they do not integrate as single units. For reasons not understood, a transgenic mouse with a human globin gene integrated at a random site will probably not have a single copy of the human gene at that site, but rather, multiple copies. Somehow, during the process of integration, microinjected gene fragments either become linked to one another or reduplicated by the DNA replication machinery, such that it is not unusual for the site of foreign gene integration to have 50 or even 100 copies of the new fragment! These long concatamers of new DNA are inherited as single units because they are all linked together into one large structure, but they of course represent many copies of the coding sequence. It should hardly be surprising under such circumstances that expression of the new gene can be at levels far higher than would be desired if the gene was injected for therapeutic or even genetic enhancement purposes. Some transgenic mice produce thousands of times more protein from their transgenes than they do from their corresponding chromosomal genes. In some livestock engineered to produce human proteins in their breast milk, the breast glands have become inflamed because the large amounts of the new protein form a sludge in the milk and block the gland. At the present time no method exists for controlling the number of copies of the transgene that ultimately become inserted at each new locus of integration. Given this lack of control, it is obviously unacceptable to introduce genes into the human germ line by this method. It should be recognized before we move on that extraordinary overexpression of transgenes is not the only problem with this aspect of pronuclear microinjection. Sometimes new genes integrate into chromosomal sites that permit no expression at all. Expression failure would also be an unacceptable situation in the human.

As if these problems weren't enough, there are disturbances of host DNA that go beyond simple insertional disruption of host genes. When the chromosomal DNA flanking the insertion site of microinjected transgenes is studied carefully, it becomes clear that integration of the foreign gene does not occur by having the embryo carefully "snip open" the chromosome and slip in the new fragment. Rather, the chromosomal DNA is often scrambled, such that sequences become interspersed with regions of the new gene. In addition, it is not unusual for the chromosome to lose as much as a million bases of DNA when the new fragment integrates. In some transgenic mice, insertion of the foreign gene has even been associated with complete breakage of the chromosome and translocation of a segment of one chromosome to a different chromosome. These disturbances, when they occur by natural causes in humans, are frequently associated with disastrous developmental abnormalities. Given all of these problems, it is obviously unthinkable to perform microinjection in humans. One of the reasons for this book, however, is to make such points clear to those who want to barge ahead with human germ line gene manipulation. Shortly after I produced the first transgenic mouse with Dr. Ruddle, I was contacted by an IVF clinic, which asked me to perform pronuclear injection on human embryos for correction of sickle cell disease!

Gene Transfer Methods That Circumvent the Problems of Pronuclear Microinjection

When these protean problems associated with pronuclear microinjection are considered in the context of human germ line gene manipulation, we realize that the approach is simply not workable. As we dream wistfully for a better approach, we can't help thinking back to targeted integration in tissue culture cells. With targeted integration, the new DNA fragment is discretely swapped for a corresponding chromosomal sequence. Exactly one copy of the new genetic material is added, and the site of insertion is perfectly controlled. If only we could perform targeted integration in the intact organism! However, as should be apparent from a review of the methods of obtaining and detecting targeted integration events, the tissue culture environment is a prerequisite. Only in this setting can large numbers of cells undergo gene transfer with selection be grown into colonies, and then be analyzed by PCR for those rare gene transfer events. Incredibly, techniques have now been developed that allow gene transfer to be performed in tissue culture, after which the genetically transduced tissue culture cells can be used to produce an intact organism! This remarkable capability was first attained in the mouse with embryonic stem cells, and has since been extended to a variety of species by cloning technology.

Embryonic stem cells, or ES cells, as we call them for short, have characteristics that you should readily be able to predict from the fact that they are used to achieve targeted integration and subsequent production of an intact animal. Obviously, these cells must be amenable to long-term maintenance in tissue culture, for time is necessary to perform DNA transfer, or "transfection," followed by selection for, and detection of, targeted integration. But also, these cells must retain developmental totipotency. That is, they must have the potential to activate any and all genes needed for complete embryonic development.

Given these latter requirements, it should not be surprising that ES cells are derived from early embryos. The way to produce these lines is to take blastocysts and place them in tissue culture instead of implanting them into the uterus. The tissue culture conditions are highly specialized to allow sustained multiplication of the few cells in the ICM that retain developmental totipotency. The cells determined to form extraembryonic parts cannot maintain themselves in tissue culture, and so, under these special culture conditions, colonies of cells emerge from the ICM and overtake a cell population that is initially a mixture of the two cell lineages in the blastocyst. Tissue culture is often stressful to cells. When a skin biopsy is performed and the excised cells are placed in tissue culture, they cannot be maintained for more than about 25 cell divisions. In addition, analysis of the chromosomes of cells cultured for extended periods often reveals major structural aberrations not entirely dissimilar from those described as resulting from integration of microinjected DNA. Despite these problems, it is possible to develop and maintain ES cell lines for long periods of time without the development of significant structural changes in the DNA. Once these cells are subjected to gene transfer and targeted integration events are detected, it is then necessary to use the cells for reconstruction of an intact animal. How is this latter step accomplished?

Recall in our earlier discussion of early development that the embryo is quite plastic: It is able to develop well when cells are lost and able to accept new cells that are added (see pp. 105–106). The way to produce mice from genetically transduced ES cells is to take these cells and simply inject them into the blastocyst cavity of another embryo. The ES cells then crawl into the ICM region of the blastocyst and pick up development where they left off at the time their own blastocyst was set in culture. It is for this reason that I often like to refer to ES cells as "blastocysts in suspended animation." Of course, when the ES cells enter the ICM of the blastocyst into which they are inserted, they must share that site with the ICM cells that were already present. What happens in this situation is quite similar to the result seen when embryos are aggregated at the eight-cell stage as shown earlier (see page 106). The mouse that develops is a composite or genetic mosaic, with cells derived from both the ES line and the host ICM. The mice look mosaic, like the one in Figure 32 on page 106, and successful colonization of the ICM by the injected ES cells is identified by using coat color markers in same way as is done for embryo aggregation.

The genetic mosaicism manifest as a mottled coat color pattern is of course present in all internal organs as well. It should be readily appreciated, therefore, that not all cells of these "founder" mosaic animals have altered genes. Only the cells with the coat color markers of the ES cell line have genes inserted when the ES cells were in culture. These mosaic mice may therefore not exhibit the full effects of gene transfer, because only some of their cells have been given new genetic material. To ensure that all cells of the animal have been genetically engineered, it is necessary to breed the mosaic mouse and hope that a sperm derived from the ES cell line, and carrying the parental chromosome that has received the new DNA, fertilizes an egg to produce a mouse, all of whose cells carry the new gene inserted in cell culture. You may perhaps wonder why I say "sperm" instead of "sperm or eggs" when I refer to the developmental fate of the ES cells in the mosaic embryo. I'm not a male chauvinist! The fact is that ES cell technology works more easily when XY ES cells are used and the initial mosaic mouse is a male. Not only can males produce many more offspring and provide a better opportunity to find offspring derived from the ES cell component of the mosaic parent, but XY ES cells allow insertion of genes onto the Y chromosome, which cannot be done if XX ES cell lines are used. Another important convenience of XY ES cells is that when they are inserted into a "female" XX blastocyst, they induce male development and are able to produce germ cells (recall that mosaics with XX and XY cells usually become males). XX ES cells in a XY blastocyst would be unable to make sperm in the resultant male mosaic mouse.

To obtain ES cell-derived sperm it is of course necessary that during development of the mosaic embryo some of the cells selected to become germ cells at the step of determination are derived at least in part from the ES component of the mosaic embryo. If all of the germ cells are derived from the host embryo component, none of the offspring of the mosaic mouse will have the new gene. This kind of bad luck is unusual, but there are other ways in which transfected ES cells may be unable to produce sperm. As stated earlier, tissue culture can be stressful to cells and ES cells can undergo changes in tissue culture that disturb the process of meiosis when these cells later become germ cells and attempt to differentiate into sperm. These subtle changes are not visible to the naked eye. It certainly can be disappointing to go through a complete targeted gene transfer procedure only to find out when the mosaic mouse is finally produced that the ES cells were damaged and unable to make sperm. I make this point here to illustrate some of the hazards of ES cell gene transfer, but also to underscore the point that ES cells that appear normal may not in fact be normal. The gold standard for a normal ES cell line is its ability to produce sperm. This point should be kept in mind when we discuss human ES cells later.

Although ES cells can clearly provide a method of obtaining targeted gene transfer events that can be transmitted to offspring through the germ line, it does have significant limitations. For reasons yet unclear, it has not yet been possible to produce developmentally totipotent ES cells that are suitable for gene transfer in any species other than the mouse. When attempts to make ES cells fail in other species, the cells often will not grow for long enough periods to allow for gene transfer or, when returned to blastocysts, will not take up residence in the ICM and contribute to the embryo. Even in the mouse it is quite difficult to make ES cells. Only some highly inbred strains of mice produce embryos suitable for establishment of ES lines. An appreciation of the great difficulty of making developmentally totipotent ES cells should give one pause when hearing press announcements that bona fide human or monkey ES cell lines have been made.

Although ES cell technology is fickle and temperamental, it is certainly possible, in expert hands, to obtain targeted gene integration in mouse ES cells and produce offspring derived from those cells. This being the case, what kinds of things are done with targeted integration?

The most common application of targeted integration technology is its use for "knocking out" genes. One can introduce a new gene into ES cells that cannot be expressed and swap it for a chromosomal gene that is expressed. The result is a mutation that negates gene function. An easy way to obtain such "knockout mutations" is to put the Neo^R gene into a place that interrupts the coding sequence of the gene to be replaced. For example, if the Neo^R gene replaced the ATG codon that initiates translation, the new gene either would fail to be translated or would insert amino acids from the Neo^R protein into the protein that was encoded by the replaced gene. In either circumstance, the function of the host gene would be destroyed by the swapping event.

Why would we want to knock out a gene? The ability to eliminate a specific gene from an animal can provide information on developmental mechanisms that would be extremely difficult to obtain by any other method. Consider, for example, the rejection of foreign tissue after organ transplantation. The mechanism whereby foreign cells are identified and eliminated is a key element of the immune response. Moreover, the same membrane proteins that provide an immunologic signature to every organism are complexed with viral proteins when a viral infection occurs. It is the complex of these proteins with viral proteins that allows the immune system to recognize and destroy infected cells. The surface proteins that perform this dual function are present on every cell and are presumed to play a key role in development as well as function of the immune system. They are anchored at the cell surface by another molecule, β_2-microglobulin. Without β_2-microglobulin, the signaling proteins, called major histocompatibility proteins, cannot be stably anchored on the cell surface.

To investigate the importance of these histocompatibility proteins in development and function of the immune system, two independent laboratories knocked out the β_2-microglobulin gene in ES cells and produced offspring carrying this knockout from the mosaic animal. Male and female offspring carrying the knockout mutation on one chromosome were then crossed to produce mice with the destroyed gene on both parental chromosomes. This breeding strategy is identical in concept to the production of a genetic disease from a recessive mutation such as the Tay–Sachs mutation (see page 82 for a review): One of four offspring of two mutation carriers will be homozygous for the mutation. Through this strategy mice were made that had no β_2-microglobulin and that, consequently, were unable to place the histocompatibility proteins on the surfaces of their cells. Remarkably, these animals developed fairly normally and survived. Despite some relatively minor disturbances in their immune systems, they functioned almost normally. Interestingly, these mice would be universal transplantation donors, because the proteins recognized for tissue rejection are absent from their cells' surfaces. The implications of a genetic engineering strategy that would generate tissues that could be accepted by a completely unrelated recipient need not be emphasized further.

Another interesting example of the use of knockout technology involves removal of genes whose protein products help protect the organism from cancer. One such "tumor suppressor gene," called p53, is mutated and nonfunctional in many human cancers, a finding that suggests that this protein plays an important role in protection against malignancies. Investigation of p53 function has shown that the protein indeed does protect against uncontrolled cell division, which is of course a key feature of cancers. p53 not only suppresses cell division but activates mechanisms that induce cells to commit suicide if they try to replicate their DNA inappropriately. By "inappropriately" we refer to the fact that cells must replicate their DNA to divide and that in tissues where development is complete and cell division is no longer taking place, DNA replication, as would occur if a cell became cancerous, would be inappropriate.

When the p53 mutation is knocked out, mice carrying the deficiency are just about normal, but of course each cell of these animals has only one intact parental copy of the gene. Therefore, if a chance mutation occurs in the single remaining copy, the cell is left without p53 and it will not commit suicide if it becomes a cancer cell. When these mice are given compounds that cause DNA mutations, they develop cancers with a far higher level of sensitivity than normal mice. These animals thus allow for highly sensitive testing of compounds for carcinogenic potential. When carriers of the p53 knockout mutation are bred to produce ¼ animals without any p53, they are born and are able to reproduce but develop multiple tumors. These are just two examples of the enormous power and utility of knockout technology.

In addition to providing an opportunity to knock out genes, targeted integration in ES cells can allow substitution of one allele of a gene for another. In the case of sickle globin, for example, one could use recombinant DNA technology to introduce a point mutation into the mouse β-globin gene that corresponded to the human sickle gene and then perform a targeted integration and swap the mutated gene for the normal copy in the chromosome. Once mosaic mice are made that transmit the mutation from the ES line to offspring and the offspring are bred to one another to produce mice homozygous for the replacement, it is theoretically possible to create a mouse with "mouse" sickle cell anemia. Obviously, it is also possible to replace defective genes with normal counterparts.

It should be clear from this discussion that targeted integration in ES cells offers a better prospect for safe and effective engineering of the human germ line than does pronuclear microinjection. What's wrong with it, then?

Even a glance at this technology should make it immediately obvious that it is not a viable approach to human germ line gene transfer. To apply this approach to humans, it would first be necessary to establish ES cell lines in this species. We've all read about the development of human ES cell lines in the press and the decision of the Bush administration to allow federal funding of research on extant lines but not for establishment of new lines. All of the discussion surrounding this controversy has assumed that the existing human ES cell lines, established from frozen human embryos generated in IVF but later abandoned or donated to research, are developmentally totipotent. But, as pointed out in our discussion of mouse ES cell lines, there is no way of knowing that ES lines are completely normal unless they are inserted back into embryos and demonstrated to produce all cells of an adult organism, including germ cells. It is not possible to perform such a test in humans, so it is not possible to document that any of the human ES cell lines could actually be used for germ line gene targeting. Given the great difficulty in establishing ES cell technology in species other than the mouse, the likelihood is that the human lines are not in fact developmentally totipotent.

Another problem, of course, is that gene targeting in ES cells is followed by insertion of the cells into a blastocyst and production of a genetically mosaic founder animal. To perform such a procedure in the human would obviously be unacceptable. We cannot produce genetically mosaic humans for the purpose of breeding them to obtain germ line transmission of their ES cell component. Yet another problem is that an ES cell line, even if it could be used for human genetic engineering, could only be used once. If a single ES cell line were used to produce more than one individual, then of course all of the individuals produced would be, with the exception of the single targeted genetic locus, identical. The use of ES cells in this way would thus be a form of reproductive cloning. If a family were to desire gene transfer for correction of a genetic disease in a future child, it would be necessary for them to produce embryos, establish their own ES lines from those embryos, perform gene transfer, and insert the ES lines into a blastocyst from another of their embryos. It might then be possible to ensure that the ES cells take over the ICM of the blastocyst completely, such that the child born would have only cells derived from the ES cell line. If the family carried a recessive genetic disease such as Tay–Sachs disease, replacement of even one of the mutated genes with a normal counterpart would be therapeutic. However, this is a very heroic and indirect strategy for treating genetic disease that cannot be justified when far more straightforward approaches exist. And, as stated before, there would be no way of knowing that any ES cell line established from a human embryo would have developmental totipotency.

Given these problems, it might perhaps be of interest to briefly discuss the high interest in human ES cell technology. If it's not good for germ line genetic modification, what is it good for? Historically, mouse cell lines very similar to ES cells, called teratoma cells, have exhibited some remarkable features. These cells can be grown in a tissue culture dish and retain the potential to produce a variety of specialized cell types. Under certain conditions, these undifferentiated teratoma cells can produce nervous tissue, muscle cells, and skin cells in culture. If returned to an animal and allowed to develop as a tumor, the tumors can produce hair and even teeth! The potential for differentiation of cultured cells into myriad specialized cell types in the tissue culture environment is presumably retained by ES cells. This being the case, ES cells may provide an opportunity to control cell differentiation and allow the researcher to induce formation of any specific cell type in a tissue culture dish. Thus the neurons that degenerate in Parkinson disease or Alzheimer disease might be induced to form in unlimited quantities in a tissue culture dish and then used in transplantation procedures to replace lost cells in humans with such diseases. This strategy could extend to any number of human disease states such as bone marrow disorders or loss of heart cells after a heart attack. As we know already from the gene knockout of β_2-microglobulin, it is also possible to engineer these cells such that they can be accepted as a graft by any recipient. In addition, attaining the ability to control the process of cell differentiation would clearly be accompanied by a far deeper knowledge of the processes of determination and differentiation that are keystones to successful embryonic development. Given the difficulties encountered from efforts to produce animal ES cell lines, this human research clearly depends for its advancement on the creation and characterization of many new lines, thus the controversy over the Bush administration decision not to allow funding for creation of new human ES cell lines.

Cloning and Germ Line Gene Transfer

The previous discussion of ES cells as germ line gene transfer vectors underscores the need to circumvent the step where a genetic mosaic is initially produced. Cloning by nuclear transfer provides this opportunity.

From a conceptual standpoint, cloning is not an elaborate process. Cloning takes advantage of the fact that every cell in the body has two complete copies of every gene (except, of course, genes on the X chromosome in males). One simply takes the nucleus of an adult cell, returns it to an egg, and allows the egg to undergo development. Of course, it is absolutely necessary for the genes already in the egg to be removed before transfer of the adult nucleus. This can be done by inserting a microneedle into the egg near the first polar body. At the time the egg is ovulated, the first meiotic division has been completed and the chromosomes are lined up in preparation for the second meiotic division, which takes place after fertilization (for a review of this chronology, see pp.

74–75). Although the chromosomes are not surrounded by a membrane, their position is so predictable that their removal is not difficult in the hands of an experienced person with suitable equipment. It is also important that the egg be "activated" so it will begin dividing and initiate embryonic development. Under normal circumstances activation is accomplished by entry of the sperm. However, when an adult nucleus with all genes is transferred to an egg fertilization is circumvented, so other methods must be used to activate the egg and induce it to begin developing. This latter requirement is rather easily satisfied by a number of procedures, some of which occur as part of the nuclear transfer itself. Nuclei can be put into enucleated eggs by simply fusing an entire cell to the egg with electrical shock or by sucking the nucleus out of an adult cell with a microneedle, inserting the microneedle into the egg, and expelling the adult nucleus into the cytoplasm. These kinds of manipulations often activate development, and when they fail several simple backup procedures are available to ensure egg activation.

A longstanding line of investigation in biology has entailed formal efforts to prove that all cells have all the genes. Successful cloning of an organism from an adult cell constitutes formal proof of this assumption. If complete development can be achieved with the nucleus of an adult cell, the adult cell must have all of the genetic information. For decades efforts were made to clone amphibians, with only partial success: Nuclei transferred from adult frogs to enucleated eggs led to initiation of development, but the embryo arrested at a fairly early stage. Serial transfer of nuclei from these arrested embryos back to eggs eventually led to production of tadpoles from nuclei of adult skin cells. The presumed reason for the difficulty in achieving full development from an adult nucleus was the imperative that genes be shut down permanently as the differentiation of cells leads to progressively restricted repertoires of gene expression (for a review of this imperative, see pp. 90–91). How could the nucleus of an adult cell, which has long since silenced the genes needed to direct the early steps in development that follow fertilization, be expected to reactivate the silenced genes quickly enough to ensure proper expression when the egg has been activated for only a few hours? Given the effort made to ensure that silenced genes are not inappropriately reactivated, it hardly surprised anyone that nuclear transfer in frogs required serial nuclear transfer before significant progress toward production of an adult animal could be attained. Of course, the difficulty with frog cloning was presumed to be less than would be encountered with a mammal, in which embryonic development was presumed to rely on more complex regulation of larger numbers of genes.

It is for these reasons that the cloning of Dolly the sheep was so astounding to the scientific community. Dolly was produced by replacing the chromosomes of a fertilized egg with the nucleus obtained from a cell in the adult breast gland. This nucleus was able to quickly activate the genes needed to orchestrate early embryonic development and, of course, execute the program of gene expression then required for completion of development to birth and adulthood. Although cloning offers many new opportunities and strategies for genetic engineering, it is my view that its most important contribution to science will always be the simple fact that it works.

If an organism can be produced by transferring the nucleus of a breast gland cell into a genetically "empty" egg, then it should also be possible to use cultured cells as nuclear donors. Indeed, as cloning has advanced, this has proved to be feasible in several cloned species. If cultured cells can be used as nuclear donors, they can of course also first be subjected to gene transfer. The gene transfer protocol could simply involve electroporation of new genetic material followed by selection or could involve a gene targeting protocol. Thus it is theoretically possible to perform targeted integration in tissue culture and to use the nuclei from the cells with targeted integration for cloning. When cloning is done, the intermediate step of producing a mosaic animal, as happens with ES cells, is avoided. We can now easily see how a child dying of Tay–Sachs disease could have a skin biopsy with culture of the excised cells followed by a gene transfer procedure in which targeted integration is used to replace one of the defective Tay–Sachs genes with a normal copy of the gene. These cells could then be used for nuclear transfer, and a baby could be born that was genetically identical to the dying child except for the fact that it would only be a carrier of the recessive Tay–Sachs mutation. The cloned child would therefore not have any symptoms of a genetic disease. This kind of genetic manipulation strategy illustrates the incredible potential of cloning as a gene transfer tool. By the way, if you're on your toes, you'll protest this gene replacement strategy as inadequate for humans because the procedure introduces a new gene, the Neo^R gene, into the cloned individual (if you're not sure why this is so, review the description of targeted integration on p. 128). Because addition of this bacterial gene to the developing human embryo could cause developmental problems, it would be essential for gene replacement procedures to include a mechanism whereby the Neo^R gene is removed. Although methods for removing the Neo^R gene are slightly more involved than the targeted integration strategy outlined earlier, they have been developed and they work reliably. Thus we don't have to worry about inserting any unwanted genetic material when we do gene replacement.

In theory, cloning could be much more efficient as a method of germ line gene modification, because the genetic alterations can be introduced in tissue culture before producing the clone. When gene transfer and related modifications are performed in tissue culture first, all of the tissue culture cells that survive drug selection are genetically transduced. Therefore, every nucleus used to produce a clone has the desired genetic changes and 100% of cloned organisms would also have those changes.

How efficient and effective is cloning in practice? To provide some perspective on the state of the art of cloning technology at the present time, I have summarized data from publications that report successful cloning of sheep, goats, cattle, mice, cats, and pigs in Table 2. These publications were selected because they exemplify the exploratory phase of cloning research in each of these species, which corresponds to the phase we would find ourselves in should we attempt to clone humans. In addition, the data were excerpted in such a way as to provide the fairest and most objective representation of the current state of the art, as well as to represent the phases of the cloning process that are most relevant to the clinical use of cloning. A comparison of the first two columns gives a indication of the success rate for performing the nuclear transplantation procedure. The first column shows the total number of eggs subjected to removal of the chromosomes followed by insertion of a donor nucleus, and the second column shows the number of healthy embryos that were transferred after nuclear transplantation in an attempt to achieve pregnancy. In general, all healthy embryos created are transferred, but in the study with sheep more healthy embryos were created than could be transferred. Accordingly, the

Table 2. Examples of cloning results for six species

Species	# Eggs manipulated	# Embryos Transferred	Births	Problems
Goat	285	112	3	Late fetal loss
Sheep	834	202 created, 156 transferred	5	Fetal loss 10 times normal
Mouse	2468	385	17, 10 survived	7/17 newborn or infant deaths
Pig	1368	401	5	Late fetal loss
Cow	270	33	4, 3 survived	Fetal & newborn death, multiple anomalies
Cat	188	87	1	Late fetal loss

23% of eggs manipulated produce embryos suitable for transfer
2.2% of transferred embryos yield surviving young

number created is shown in order not to underrepresent the success rate for creating healthy embryos. A summary of the six studies shows that about 1/5th of the eggs used for cloning survive to produce embryos that have significant potential for full-term development. The next column shows the number of births obtained after transfer of the healthy embryos. The birth rate per embryo transferred averages about 2% over the six studies. Thus about 0.4%, or 4 of every 1000, eggs used for cloning develop into live young. Recall that one of the problems with microinjection is that only about 1/5th of injected embryos are born with the new genes integrated. By comparison, cloning is 50 times less efficient in this respect. Remember that the frequency of gene integration in animals born after cloning is 100%, because the gene transfer is performed in tissue culture before nuclear transfer and only nuclei with integrated genes are used in the cloning process.

The final column of this table lists "problems" with the cloning process from the perspective of development after implantation. A remarkable feature of nearly all cloning studies in animals is that many of the pregnancies established are subsequently lost. Embryos that appear normal at the time of transfer to the uterus have disturbances of gene regulation that are not readily apparent by simple visual inspection of the embryos. As a result, these embryos implant and develop to quite advanced stages but are aborted before birth. Note that in the sheep study the rate of such pregnancy losses was 10 times higher when cloning was performed than it is when pregnancies are established by normal means. In mice the loss occurs shortly after birth, with nearly half of the animals dying within a week of birth. The study with cows gives some indication of why the fetuses and newborns die. As with the mice, one of four newborns in this study died shortly after birth. When this animal was examined internally, it had multiple abnormalities, mainly involving the heart and blood vessels. We may interpret these problems of late fetal and early postnatal development as pleiotropic effects of disordered gene regulation occurring probably as far back as preimplantation development.

A remarkable study on cloned mice has provided direct evidence that cloned animals have abnormal regulation of many genes. In this study, a number of genes with known patterns of expression in normal mice were examined in cloned mice. Remarkably, a large number of genes were improperly expressed. So profound were these abnormalities that the authors of this study expressed surprise that the animals were even alive! In addition to these molecular demonstrations of abnormal gene regulation in cloned mice, some overt phenotypic abnormalities such as gross obesity have been observed.

What these studies tell us, then, is that cloning is very inefficient and that, even when it does work, the animals born are abnormal. Moreover, when cloning fails, the failure frequently manifests not as the failure to achieve pregnancy but as the loss of advanced pregnancies or the death of newborns.

We will return to the discussion of cloning when the procedural options for modification of the human germ line are discussed from the clinical perspective. Before leaving this subject, however, two more points should be made. First, it is important to distinguish cloning procedures that modify the germ line from those that do not. Consider the case of a man with acute leukemia, a potentially fatal cancer of blood-forming cells. Let us suppose that, on the assumption that the technology was available, such an individual hired a woman to donate eggs, used a nucleus from one of his skin cells to perform nuclear transfer into that egg, and created an embryo that was then induced to become an ES cell line. The ES cells were then induced in culture to become bone marrow cells. The leukemia patient then underwent chemotherapy to destroy all of his resident bone marrow, which included both his normal cells and the leukemia cells, and used the bone marrow cells from the culture dish to repopulate his bone marrow with normal, nonleukemic bone marrow cells. Of course, he would not reject the bone marrow "graft" from the tissue culture dish because the bone marrow cells in that dish, having been derived from one of his own cells, would have the same histocompatibility genes as he did. This kind of strategy would not result in germ line alterations, even if new tumor suppressor genes were transferred into the ES cells before their use in the graft. The patient would have genetically modified cells in his body, but these cells would not produce sperm.

However, any cloning procedure that leads to birth of a baby, so-called "reproductive cloning," should be considered a germ line modification procedure. Production of multiple genetically identical individuals after nuclear transfer constitutes a level of control of the genes in the offspring that should be regarded as a germ line manipulation, even if genetic changes are introduced into the donor cells before nuclear transfer. Because reproductive cloning would reflect this new measure of control of the genotype of offspring, we will consider any such procedure, even if it produces a single child, to be a form of germ line modification.

Another important feature of cloning that should be appreciated is that two or more genetically identical individuals produced by nuclear transfer are slightly less alike genetically than identical twins. When identical twins are produced either naturally or by embryo splitting (see p. 104), they are both derived from the same fertilized egg. In contrast, two "twins" produced by transfer of adult nuclei into two separate eggs are not derived from the same egg. What difference does this make? It turns out that mitochondria, the power plants of the cell (see Figure 1), have a very small number of genes, and some sequence variability exists between the mitochondrial genes of different individuals. Therefore, two different eggs may not have the same mitochondrial DNA sequences, and two cloned children that developed after nuclear transfer into these eggs would therefore be slightly different with respect to their mitochondrial genes. Although this point is a relatively minor one, it is useful to remember when evaluating the potential of cloning to produce an individual with a desired phenotype. If a specific phenotype is not reproducible in identical twins, it almost certainly will not be reproducible in clones.

Other Methods of Germ Line Gene Transfer

Another approach to modifying the germ line entails infection of eggs or embryos with modified viruses. As mentioned earlier, viruses cannot replicate independently. Viruses that infect mammals enter the cells and then recruit the resources of the infected cell to replicate their own genetic material. Once replicated many times over, the genetic material of viruses is "packaged" into protein complexes that protect it during transit from one cell to the next. The information for the packaging proteins is in the genetic code of the virus, but the mRNA for these proteins is translated with the ribosomes of the infected cell. Once the virus has replicated its genetic material and produced a sufficient quantity of packaging proteins, the viral genomes are packaged for release from the cell. To ensure proper and efficient assembly of the packaging proteins around the viral genes, the genetic material also has a packaging signal—a group of bases that directs assembly of the intact virus from the genetic material and the viral proteins.

Viruses vary in the type and behavior of their genetic material. Some viruses store their genetic information as RNA instead of DNA. Two viruses of this type are retroviruses, which produce diseases such as feline leukemia, and lentiviruses, such as that which causes AIDS. When these viruses enter the host cell, they carry with them an enzyme that transcribes DNA from RNA, the reverse of the process that the host cell uses to make RNA from its DNA. This "reverse transcriptase" makes a DNA molecule that efficiently integrates into the host cell chromosome and then makes more viral RNA for packaging into mature virus particles or for translation into packaging proteins.

Genetic engineers have performed the following trick to produce "recombinant" retroviruses and lentiviruses. The recombinant viruses can carry mammalian genes such as globin and can efficiently infect cells. However, the genes for the packaging proteins are removed. Thus they cannot replicate, and exposure of mammalian cells to these agents does not lead to a productive infection resulting in release of infectious particles that enter previously unexposed cells. In addition, these modified viruses have their nefarious genes, which may cause cancer or AIDS, removed. This elegant engineering of recombinant viruses is performed as follows.

First, DNA copies of the viral genetic material are modified so that the bad genes and packaging protein genes are removed and the gene desired for transfer is inserted. The viral DNA sequences required for DNA replication are also deleted. Importantly, however, the signals on the DNA that direct packaging of the viral genomes into viral particles are still present. These DNA molecules are introduced by electroporation into special tissue culture cells. The cultured cells are special because, before receiving the recombinant virus DNA, they are first transfected with recombinant DNA plasmids that encode the virus packaging proteins. These cells cannot produce viruses because they contain only the genes that make the virus packaging proteins; they have none of the other information from the virus needed for packaging or replication. However, these cells are filled with packaging proteins. When the recombinant virus-derived DNA is inserted into the cells, they are able to assemble into virus particles because they still have the packaging signals. Although these recombinant molecules lack the genes for the necessary proteins, those proteins are supplied by the transfected cells, which thus "help" the recombinant genomes to package. In this way the recombinant virus genomes, incapable of replicating but carrying virus packaging signals and the donor genes of interest, are produced in large quantities, after which they can be harvested and used to infect other cells, including eggs and embryos.

When recombinant retroviruses or lentiviruses are exposed to unfertilized eggs or early embryos (if you're alert you'll realize that the embryos and eggs must first have the zona pellucida removed!) they infect, release their genetic material, and integrate that material into a host chromosome. This process results in transfer of a new gene or genes into the germ line. Of course, some viral sequences—packaging signals and sequences that help the viral DNA integrate—go along with the new gene.

Germ line gene insertion using recombinant viruses is very feasible technically, and retroviruses were in fact used to produce the first transgenic monkey.

The gene transferred into that monkey was from a jellyfish! However, recombinant viruses are not commonly used for transgenic experiments in part because they are limited with regard to the size of the donor gene that they can carry. If recombinant virus genomes are made too large by insertion of a large gene, they do not package properly. Another problem is that the new genes often do not express stably and at high levels. Finally, these methods do lead to integration of unwanted viral genetic material along with the desired gene, and no control over the site of integration can be exerted. Nonetheless, it is important to be aware of this gene transfer methodology because it does work and further improvements in it may eventually lead to an acceptable method of human germ line gene insertion.

With this information in hand, we are now ready to examine germ line gene transfer as it applies to humans. The relevant knowledge of developmental biology, molecular biology, genetics, and gene transfer technology is now in hand. However, now we must step out of the arena of animal experimentation and into the realm of clinical therapy. To evaluate the feasibility and ethical acceptability of extending these procedures to humans, we will have to review the basic principles governing the proper practice of medicine.

PART II

PART II

8

INTRODUCTION TO THE ETHICS OF REPRODUCTIVE GENETIC TECHNOLOGIES

Physician do no harm.
—From the Hippocratic oath

I'll never forget the moment I first set eyes on her. When I walked in, she looked up with an expression of complete trust. Her large eyes peered shyly but confidently out of her lovely face, with its soft lines, skin smooth as alabaster, and pretty little nose. She was undressed save for a pair of underwear and a bra. She glowed with the fresh look of a child, but at sixteen, she also had the alluring curves of a grown woman. Although her brow conveyed a look of self-confidence and equanimity, it was her lovely mouth, with its full lips trembling slightly as she spoke, that gave away the self-consciousness she felt as she presented to me, to do with as I wished, her naked body. I asked her to remove her bra, and she did so in a careful and measured way, but without hesitation. Her breasts were, smooth, symmetrical, and full. She leaned toward me, chin up, so that I could touch her without hindrance. She was just a few inches away. I could smell her perfume. Her breathing was rapid and shallow in anticipation of my touch. I reached forward, but then stopped. I just couldn't do it.

After all, I'd never been in a situation like this. Here I was, a Jewish medical student in his early thirties faced with this inner-city sixteen-year-old whom I had never met before, who had come to Pediatrics clinic for a preschool physical exam. Given the disarming circumstances, I just couldn't bring myself to palpate her breasts for lumps. Instead, given the fact that breast cancer is extremely unlikely in a girl her age, I advised her to examine herself when she showered and explained how that was done. I also took a look at her breasts to make sure there were none of the obvious signs of problems, such as dimpling of the skin, that I had been taught to look out for in my lectures. The symmetry of her breasts was one sign that no serious disease was present.

After the young lady departed, I reported back to my instructor and explained what happened. His response was as follows: "I see, so you decided to place the responsibility for detecting an early breast cancer on your patient rather than taking it on yourself. Do you think it likely that her skills at the breast exam are going to be anywhere near as good as yours? Do you really think that, as her doctor, you should have left this responsibility to her?" On hearing this I bolted from my chair and ran toward the door. There was a good chance I could catch her and bring her back for a proper breast examination. My mentor stopped me, however. He explained that her chances of having a malignancy were indeed quite low. But he told me that in the future I should consider my position as a physician more carefully.

Of the many reasons I recall this incident so clearly, the one that induces me to relate it here are the remarks made by my teaching physician. His comments highlight some of the many unusual rules governing the patient-doctor relationship—rules without which medical care cannot be given satisfactorily. When a patient hires a physician to provide him or her with medical care, he allows the doctor, who may be a complete stranger, to inquire about the most personal and private aspects of his life and to place his or her hands and instruments on his body without restriction. The physician would never be able to take such liberties were it not for an unstated mutual understanding between patient and doctor. This understanding includes the assumption that what transpires between patient and doctor will be held strictly confidential unless the patient desires that the confidence be broken. When a patient resolves to reveal the most private aspects of his or her physical, emotional, and social life, he does so on the assumption that no other individual has a right to the information. If the patient cannot trust the physician to maintain this confidence, the physician is unlikely to receive the full and accurate information needed to develop the best treatment plan. Therefore, a relationship of trust is struck. The patient will tell or show the physician anything and trust the physician to use that information for the sole purpose of preserving, protecting ,and defending the patient's most important assets: physical and mental health.

Implicit in the covenant between patient and doctor is the assumption that the doctor's overriding priority is optimizing the patient's approach to his or her medical problem. The days of Marcus Welby, Dr. Kildare, and Ben Casey, when the family doctor imposed his personal or philosophical views by recommending treatment plans that conformed to his own social views and mores, are gone. Today, the doctor is hired to serve the patient's needs, not to be judgmental of the patient's attitudes or way of life. When treatment options are available, the patient is entitled to an honest and dispassionate recitation of them and he or she expects to make the final decision regarding the course of action to be taken. If a patient elects a treatment option that violates the doctor's religious beliefs, social attitudes, or notions of what is proper medical care, the course of action expected is not that the physician will impose some other treatment plan, but that he/she will instead decline to treat the patient and provide a referral. Because even the most intelligent and educated patients often do not know much about medicine, the doctor is expected to provide information required for intelligent decision making in a completely unbiased and objective manner. Thus the doctor has a solemn responsibility to avoid having his or her own personal preferences influence the manner in which information is imparted to the patient. This is especially true in light of the position doctors have traditionally occupied in our society. Doctors have historically been like Dr. Welby—viewed as authorities whose judgments and opinions of one's personal conduct were important. For this reason, doctors must be careful not to communicate approval or disapproval of patients' decisions; rather, they should dispassionately provide their best professional opinion of the medical advisability and feasibility of those decisions.

When a doctor presents treatment options to a patient, many aspects of the interaction must be taken into consideration. We will discuss these in more detail in a moment. First, however, there are other overarching issues which should be mentioned briefly. The first of these is the understanding that the physician, enabled by the information developed from interaction with the patient, will do everything possible to realize the patient's treatment objectives as holistically and effectively as possible. If a terminally ill patient requests that his organs be donated for treatment of others or for research, it is not enough for the physician to simply administer proper terminal care. That patient is entitled to an assurance that his wishes for organ or tissue donation will be fulfilled. Therefore, treatment is not restricted to addressing specific and immediate medical problems; it extends to a holistic delivery of care that addresses all aspects of the situation. The emotional condition of the patient is every bit as important to address as the physical condition. If a patient loses confidence that the physician is addressing all of his health-related problems, the relationship of trust is undermined, the patient becomes less willing to share information, and delivery of care is compromised. It is this commitment implicit in the covenant between patient and doctor that I failed to honor when I didn't do that breast exam. This young patient was sacrificing her privacy in exchange for the best possible evaluation of her state of health, and I failed to provide the best.

Another important but somewhat more challenging principle to address is the notion that medical care is voluntary. Historically, people have never been obligated to seek medical treatment for any condition, and this guiding principle pretty much prevails today. If the breadwinner in a large family develops an easily treatable but otherwise certainly fatal medical problem, such as a melanoma on the forearm (melanoma is a serious form of skin cancer that is easily treated in its early stages by surgical removal), we would all like to see that individual elect to have the melanoma removed. Yet, even if her death could leave a family destitute and dependent on government financial assistance, there is no law obligating her to sign a consent form for the minor surgery. This protection against being forced to receive medical treatment extends even to situations in which society has a compelling interest in the individual's health and survival. Even the President of the United States can refuse to have a melanoma removed, opting instead to die in office. This notion that medical care is a matter of choice is based in the understanding that, of all our possessions and assets, none should be under more sovereign ownership and control than our own bodies. If the state can intervene and tell us what to do with our leg or our spleen, our personal freedom is profoundly compromised.

There are circumstances in which sovereignty over the body is not immune to societal intervention, and these circumstances usually arise when medical decisions are made by one party for another. Parents are charged with the responsibility of caring for their children, and they are entitled to raise their children in a manner consistent with their own philosophies and religious beliefs. However, this authority can be overridden by the state if it is believed that a medical treatment decision endangers the welfare of a child. The obvious example is when a minor child from a family that is opposed to any form of medical intervention, including blood transfusion, suffers an accidental injury that doctors determine will be fatal if a transfusion is not given. In these cases, doctors or medical institutions have been successful in taking over this decision making responsibility and ordering a transfusion. Similarly, the death of a child who becomes ill and is not brought for medical attention in a timely manner can result in prosecution of the parents, even though they may not "believe in doctors." Although these legal restrictions on parental decision making regarding their dependents are widely accepted and rationalized on the basis that society has too great an interest in its children to allow them to be endangered by poor judgment of their parents, one should appreciate that the laws validating such restrictions can cause problems. A transfusion for a child who is bleeding to death represents a straightforward medical crisis, but what about health problems where the outcome is not so certain and/or the treatment not so established? Removal of parental authority over the manner in which children are reared and cared for might then be viewed as unfairly officious and meddlesome. Even in the straightforward transfusion situation, imagine what would happen if a transfusion was mandated by a court order for a child of parents who opposed transfusion and then the wrong blood type was given, killing the child from a transfusion reaction? Although there are accepted legal arguments to deal with such unusual circumstances, it would be hard to make those arguments convincing for the parents, who had believed all along that the outcome of an illness or injury should remain within the province of God. It is also interesting that, although failure to seek timely medical attention for a child can lead to legal prosecution and punishment, there is no restriction of a family' choice to live many miles from any doctor or hospital, where a child could become sick and die because a doctor cannot be reached in time. These inconsistencies and difficulties serve to illustrate the problems that arise when the principle that medical care is optional is even slightly weakened.

Other situations in which medical care can be insisted upon involve issues of public health. Vaccination is required for several infectious diseases for all children who attend the public schools. It is also possible, of course, to quarantine people when an infection breaks out that threatens to become an epidemic. In situations such as these, logical consistency is far more easily maintained because there are numerous legal precedents for curtailing some individual freedoms in the interest of public safety. However, it is quite important that the fundamental right of an individual to control what is done with his or her own body is very highly valued and is protected vigilantly by the medical consent process.

When we discuss medical interventions required for engineering the human germ line, we should also keep in mind that implicit in the principle that medical care is a matter of choice is the notion that patients can discontinue treatment at their own discretion. Although it may seem obvious that therapy can always be refused at any step in the treatment process, this basic tenet of medical practice can be especially relevant when experimental procedures with ill-defined risks and no obvious benefit to the patient are undertaken.

A final general consideration involves the prospect that the physician or medical team might have something to gain from a patient's choice of a particular treatment option. Traditionally this kind of conflict of interest involved encouragement of patients to opt for treatment that was the most remunerative for the doctor. In the area of biotechnology, within the purview of which are the strategies and methodologies of germ line genetic modification, there are other potential conflicts as well. The first practitioner to clone a human being might stand to benefit financially not only from investment of venture capital in his/her biotechnology company but also from the fame he or she might acquire. Regardless of the nature of any conflict of interest, it is absolutely critical that the patient be advised when it exists and of its nature as well. The patient must then weigh this information when the physician's advice is evaluated. With this background we can now consider specific aspects of the process whereby the physician presents treatment options to the patient.

One obvious fact that has been largely overlooked in discussions of human germ line gene manipulation is that the process is a form of medical therapy. Many of the procedures required are invasive and, as such, associated with risk. They also can cause pain and discomfort, as well as emotional distress. It is improper to ignore these crucial facts when entertaining the notion of genetic manipulation. All parties involved in the venture are patients whose rights must be protected. It is within this context that we now consider some of the important issues surrounding the choice to undertake therapy and, if such a choice is made, which treatment option to select.

Assessing the Need for Any Intervention

Regardless of the kind of therapy proposed to a patient, the need for any therapy at all must be weighed. When a patient shows up in the office with acute appendicitis, the decision to treat is an easy one. Without treatment the patient could very well die, whereas with treatment, which consists in removal of the appendix under general anesthesia, there is a chance that death will occur from the anesthetic or from an error on the part of the surgeon. However, these risks are statistically miniscule compared with the risk of doing nothing. The choice is easy here: Go ahead and treat.

On the other extreme we have procedures that are completely elective, such as cosmetic surgery. A patient may wish to have a breast augmentation, liposuction, or a face lift. None of these procedures is necessary for survival, but the treatments still carry risk. Under these circumstances the decision not to treat at all must be seriously considered. This example is not intended to suggest that cosmetic surgery is not occasionally of major significance for a patient. Correction of physical abnormalities can so dramatically improve the lives of some patients that a strong impetus exists to undertake treatment. Rather, I am simply pointing out that the risk of intervention must be considered more carefully when not intervening is a viable option.

Where genetic manipulation is considered, the need for therapy varies widely. One prominent proponent of cloning has announced that he would clone himself "for fun." The therapeutic imperative here is very low, of course, but this individual is clearly prepared to endure the medical intervention in order to fulfill his desire to clone. In any case, the intervention is minimal for him—a biopsy to obtain a few cells. However, women who must donate eggs and carry pregnancies are also involved, and the risk they would undertake is substantially greater. Because it is unlikely that these participants have much to gain from "cloning for fun," it is presumed that they would be compensated financially for taking on the risk. We will discuss these more substantial risks later when cloning is compared with other forms of intervention for treatment of genetic disease. It should already be clear, however, that the decision of patients to place themselves at physical and emotional jeopardy just for money can create some ethical problems.

Genetic manipulation could be used to cure disease or reduce the risk of disease. In circumstances such as these the risks for the would-be parents are more easily justified, with the degree of justification depending on the seriousness of the condition being addressed. When therapy for disease is the objective, genetic manipulation procedures must then be compared with treatment alternatives from a variety of points of view.

Judging Procedures for Effectiveness

It is not ethically acceptable for an individual to be subjected to painful and potentially dangerous medical procedures if those procedures do not effectively address the medical problem. Although this statement seems obvious, there are various parameters of effectiveness that should be considered. A procedure that is completely curative when it works is not very effective if it works in only one of a thousand patients. Another incident from my medical school career serves to illustrate this point. One morning, our medical team was discussing a patient with newly discovered lung cancer. The attending physician pointed out that surgical removal of the tumor had the potential to be completely curative. However, he said, such a procedure would never be performed because the chances of actually removing all of the tumor and effecting a cure were prohibitively low from a statistical standpoint. Because surgery would inevitably be followed by chemotherapy or radiation, the proper course, he said, was just to give the chemotherapy and radiation and spare the patient from a very invasive surgical procedure that was likely to nothing more than reduce his breathing capacity by removing large amounts of normal lung tissue along with the tumor. When this argument was presented I protested, saying that surgery was the patient's only chance. The attending physician was impressed by my fervor but not moved by my logic.

Another parameter of effectiveness is the degree to which a given intervention approaches a complete cure. Removal of the appendix or successful resection of a cancer are curative procedures. But what about operations like coronary artery bypass? Bypass operations do nothing to address the underlying disease that leads to blockage of the coronary arteries, and even when they restore normal blood flow to the heart completely, the restoration is often only temporary. In the case of bypass surgery, extensive data exist that define the circumstances under which this costly operation that involves opening the chest and stopping the heart for several hours achieves a better clinical result than medications that increase cardiac blood flow. Faced with crippling heart disease and the possibility of a fatal heart attack in the absence of surgery, most patients afflicted with heart problems that are best addressed with bypass surgery are more than willing to undergo the procedure. However, it is important to keep in mind that a cure is not effected and that risk is involved.

When we consider issues of effectiveness in the context of germ line gene intervention, we shall see that these procedures are also likely to be quite varied in the degree to which they achieve their clinical objective. Some interventions might correct a genetic disease inherited strictly as a single-gene trait, like Huntington disease, Tay–Sachs disease, or sickle cell disease. However, other strategies designed to reduce the risk of heart disease or breast cancer might not be as effective, because they might reduce but not eliminate the disease risk. As we shall see later, use of genetic manipulation to enhance the characteristics of an individual rather than correcting a genetic disorder may have a very low degree of effectiveness. Under all of these circumstances, the proper approach will be to compare the effectiveness of genetic manipulation with that of possible treatment alternatives.

Morbidity Associated with Treatment

Another parameter of comparison between various treatment options is morbidity, or the amount of suffering or injury incurred by the treatment process itself. Often, the onerous nature of a prolonged interaction with the medical profession, invariably punctuated by poking, probing, and scanning, can be very debilitating psychologically. Most of us regard the taking of a blood sample as little more than a nuisance. However, in my medical training I have seen patients pass out when blood is drawn, and I have frequently seen hospitalized patients, exasperated at having to subject themselves to daily needle sticks, simply wake up one morning and refuse blood sampling. Thus even this simple invasion is associated with some degree of suffering.

Of course, blood drawing is one of the least difficult procedures to endure. When medical therapy becomes more complicated, patients are often required to suffer terribly for quite prolonged periods. Most surgery is associated with preoperative testing that is annoying and nerve-wracking, and nearly all surgery causes postoperative pain. Moreover, rehabilitation from many surgical procedures is often time consuming and painful as well. In some instances, surgery can also result in anatomic changes that are very difficult to bear both physically and psychologically. Major surgery for oral cancer can be disfiguring, and colon surgery resulting in placement of a colostomy bag can be very difficult for the patient. Of course, surgery is not the only arena in which medical intervention causes pain and distress that is additional to the disease process being addressed. Chemotherapy for cancer can often be extremely difficult for patients, who often develop terrible nausea and vomiting as well as debilitating anemia and disfiguring hair loss. It is not at all unusual for such patients to eventually refuse continuation of such therapy and elect to die. When a physician presents alternative therapies for any medical problem, it is critically important that all of these aspects of each therapeutic approach be described honestly and completely. The goal of medicine is not to treat a disorder and ruin the patient's life in the process.

How does this principle bear on the issue of genetic modification? In all of the public discussion of cloning and germ line gene transfer, I have never heard this issue brought up. However, in their present state of refinement, many of these interventions incur known and potentially long-term risks. In the field of somatic gene therapy, where genes are placed into some adult tissues for therapy of disease, but where no attempt is made to modify the germ line, many have suggested that the patients undergo lifetime follow-up to ensure that the gene transfer did not create unforeseen medical problems. Lifetime follow-up is a very burdensome form of morbidity, and if germ line genetic manipulation required such follow-up, its desirability as an approach to disease would be commensurately reduced. When we compare genetic modification with other interventions for some disorders later, this issue will come up.

Cost of Treatment

When presenting therapeutic options to patients, the issue of cost must come up. The costs of medical care, and reimbursement for those costs, is an exceedingly complex issue, and it is beyond the scope of this book to "straighten the hair of Medusa" by attempting to make sense of the health care reimbursement mechanism currently in place in the United States. Suffice it to say that germ line genetic manipulation procedures are not likely to be reimbursed by insurance companies in the foreseeable future. This being the case, it will be important for patients to know what such a treatment option will cost them.

In the field of assisted reproductive technologies, which are an inseparable part of all germ line gene manipulation strategies, the cost has been high and paid for largely out of pocket. The more exotic use of these technologies, in combination with recombinant DNA technology, promises to be extraordinarily costly and not reimbursed by insurance. Therefore, if a patient is considering this approach to a medical problem, he or she must be aware of the high price tag and weigh this factor against the cost of alternative therapeutic approaches.

The issue of cost is more far reaching than its potential burden to the patient. Extreme costs can make such procedures available only to the wealthy. If any of the gene transfer methodologies becomes a medically viable and attractive therapy, its limitation to the privileged few could cause dissention. Although there is ample historical precedent for "medical discrimination" based on race or economic status, the profound nature of genetic modification could intensify debate over this issue. Another factor is that the cost of health care is so burdensome to society as a whole that "luxury" procedures are increasingly frowned on as consuming an unfair share of our health care resources. Although these broader social issues are not likely to become significant in the near future, further advancements in gene transfer technology, accompanied by increasing costs for equipment and labor and a decreasing pot of money available for health care, could make these issues far more contentious than they are today.

Risk-Benefit Ratio

An issue closely related to that of effectiveness is popularly referred to as "risk-benefit ratio." It is not at all unusual for participants in medical research studies to subject themselves to procedures that carry risk while offering minimal or no benefit. These situations often arise when the participant or someone close to him has an intractable medical disorder and a novel therapeutic approach is in its early phases of testing. For example, such a study could involve escalating doses of a new medication as a test for safety of that medication. Many participants in such studies often have no chance of accruing benefit from the test, and, of course, it is medically unethical not to advise these volunteers of that fact before enrolling them in the trial. It is accordingly not unprecedented for people to submit themselves to medical interventions that carry risk without offering at least a corresponding degree of benefit. However, such situations generally arise in the research setting, where the benefit mainly consists in satisfaction on the part of the research subject that his or her participation in the study will help others. Organ donation, where a donor may give a kidney or segment of liver to another person, is another example of such a situation.

Because a reasonable balance between risk and benefit is expected when therapy is undertaken, it is important to point out that effectiveness of therapy can influence risk assessment. If there is even a small risk of morbidity or mortality from the treatment itself, this risk becomes relatively more important if therapy is less effective. The extreme example to illustrate this point is therapy that has no effectiveness whatever. Under these circumstances even a low chance of harm resulting from the treatment is significant. Thus, although we cannot consider a particular therapy to be more "risky" just because it is less effective, we may regard the risk with greater weight when treatment is elected.

The Importance of Preclinical Testing

Given the importance of the risk-benefit ratio in clinical decision making, it is important to ask how risk is assessed. To assess risks of new drugs or medical procedures, it is necessary to perform safety tests either on cells in culture or in intact animals. For some drugs suspected of causing DNA mutations, tests on bacterial cells are satisfactory. Large numbers of cells can be exposed to high amounts of the drug, and systems exist for identifying mutations quite sensitively. However, for many drugs, and certainly for manipulations that involve alteration of the genetic code in a developing embryo, cell culture systems are not adequate. When we recall the phenomenon of pleiotropism (see page 93) we realize that in developing systems an injury to a cell may not be immediately discernible but may manifest as an array of abnormalities in a wide variety of cell types once embryonic development is completed. It is for these reasons that preclinical testing in intact animals is extremely important. These issues are especially relevant to genetic modification, which can have effects on development from the time of fertilization into adult life.

How much are animals like humans? In many respects a developing mouse is much like a developing human. The fundamental steps in determination and differentiation are the same, development takes place inside a uterus, with contact between fetus and mother occurring via a placenta, and the finished product is quite similar with regard to the function and placement of major organ systems. In addition, mice and other experimental animals can develop diseases such as cancer that resemble the corresponding diseases in humans in their behavior and clinical course. The individual cells of other mammals are also quite like those of humans with regard to the housekeeping genes they require and, to a large degree, the specialized patterns of gene expression they develop as differentiation proceeds. It is quite safe to say that drugs or chemical poisons that cause birth defects in mice should be regarded as highly dangerous to a pregnant woman. Similarly, drugs that cause cancer in mice, rats, or other mammalian models are likely to be harmful to humans. For these reasons, animal testing is the mainstay approach to determining the safety and efficacy of many new compounds or procedures developed to treat human disease. Although animal "rights" activists would argue that animals are not needed for such preclinical testing, even a superficial appreciation of the differences between individual cells in culture and a multicellular, developing organisms in which cells interact both directly and at a distance and function cooperatively and in highly specialized ways to maintain the life of the colony, makes clear the necessity for preclinical testing in an intact organism. Therefore, as we develop new germ line genetic manipulations for possible use in humans, it is absolutely essential that these interventions first be studied in animals.

How much are animals unlike humans? Animal test systems are artifactual in some ways. When mice are tested for effectiveness of chemotherapy, they are frequently inoculated with a known number of tumor cells, treated at a specified time after inoculation, and studied for such parameters of tumor growth as life span. Mice used for such tests are often highly inbred. That is, for generations they have been bred brother-to-sister, such that the allelic variation becomes very low. In fact, some strains of mice, even though they have been maintained in separate colonies around the world for many years, still appear as genetically indistinguishable as identical twins. They can accept skin grafts from each other, for example (recall that genetic variability in histocompatibility genes leads to tissue rejection, as discussed on page 141).

This kind of testing is therefore not perfectly representative of malignant disease in humans. When people consult their doctors and a cancer is diagnosed, the precise cell type is not well characterized, the number of tumor cells and their exact distribution in the body are not known, the time elapsed since development of the tumor is not known, and the person with the disease is not as genetically well characterized as mice are. For these reasons, the response of a tumor to chemotherapy in the clinical setting can be quite different than in an animal test system. Animal tests of this kind are still very important, because they can determine whether a drug is active against a cancer, but the ultimate clinical outcome of treatment in humans is more difficult to predict from mouse studies. In the emerging field of somatic gene therapy, where new genes are inserted into tissues other than germ cells, such as liver or brain, for treatment of cancer, the results have generally been better in animal models than in humans. Given the knowledge of the character and stage of the disease in the animal models as compared to the human, this is not surprising. However, it should also not be depressing. That these new gene therapy strategies are workable in animals is encouraging evidence that with further development and refinement they will be effective in humans as well.

A very important step in developing animal tests is choosing the appropriate animal model. Where new therapies involving the heart and circulatory system are concerned, pigs are very good models. The pig heart is similar in size and overall function to the human heart, and complex surgical procedures can be readily modeled in this large animal. When testing drugs for their potential to cause mutations, it is more important to use an animal like a mouse, in which large numbers of offspring can be quickly produced and studied for abnormalities. When testing for alterations in advanced cognitive functions such as reasoning or memory, or when modeling degenerative diseases of the brain, non-human primates are useful models. Brain structure and neuronal circuitry in these animals more closely resemble those of humans, and the complex behaviors of which primates are capable allow for more sophisticated tests of changes in cognition.

Some kinds of human disease are very difficult to model well in any animal system. Humans can be afflicted with disorders that manifest as behavioral changes that would be very difficult to detect in an animal. These disorders are relatively subtle but are readily identified in the human population because detection systems are so sensitive. A learning disability, for example, can be detected because there exists a formal process of education. Were no attempt made to teach children complex reasoning skills that cannot be attained by any animal, learning problems would be very difficult to detect. Behavioral problems such as hyperactivity, autism, and psychosis would be similarly difficult to model in an animal. Surely we see research animals with behavioral abnormalities, but we have no way of knowing that these abnormalities have the underlying cause, or pathogenic mechanism, as any known human disorder. This reality makes an assessment of the risk that a specific intervention will cause problems with development of advanced cognitive functions very difficult. When a risk of such problems is unknown, but an understanding of the procedure leads one to suspect that it might be significant, this information must be provided to the treatment candidate. To impart such information objectively, fairly and completely, it is important that the health care provider be intimately familiar with preclinical data obtained from animal experiments. It is also important to inform the patient when preclinical tests have not provided information relevant to what is recognized as a potential risk of a procedure.

The Informed Consent Process

It should be clear from the foregoing discussion that no medical procedure should be performed without the informed consent of the participant. When a patient signs an informed consent form, he/she is averring an understanding of the procedures to be performed, the likelihood that these procedures will be successful, the financial burden, if any, the benefits likely to be accrued from the treatment, and the risk of treatment. Such forms should also document that treatment alternatives have been presented comprehensively and objectively and that the choice of therapies was made with a full understanding of those alternatives. The patient also understands that, to the extent possible, he/she may discontinue treatment at any time. Finally, especially where novel or experimental therapies are undertaken, the participant acknowledges that he/she has been made aware of any vested interest on the part of the care provider(s), whether this interest be in the category of fame or fortune.

A consent form that contains all of these provisions is a prerequisite for proceeding with treatment, but informed consent is a process that goes beyond the mere signing of a form. Patients who want to proceed with treatment are often quite willing to sign consent forms, even though they do not fully understand them. Therefore, a proper informed consent process should include documentation on the part of all parties that the contents of the form are fully understood. The practitioner may be motivated to explain the contents of the form in such a way as to "pooh-pooh" the risks and wax ebullient over the benefits of treatment, or vice versa, depending on his/her biases and verbal skills. Where novel therapies are concerned, risks are more likely to be uncertain, and a full understanding of the therapeutic regimen is also more likely to be unclear. Therefore, it is sometimes advisable for consent monitors to be present when the treatment is discussed, such that a greater degree of assurance is provided that the patient really knows what he/she is getting into. A proper informed consent process is necessary for the ethical practice of medicine for the reasons previously stated. However, from the point of view of caregivers, it also provides a degree of protection from law suits in the event the patient is harmed. It is generally felt that the patient should not have the right to sue for a bad treatment outcome if he or she understood the risks of the treatment and agreed in writing to go forward.

Even when all of these protections are in place, however, the informed consent process is imperfect. Of course, it is difficult for physicians who want to perform a procedure to be completely objective in the presentation of the options. There are also communication problems that can lead to the patient misunderstanding what is said. My mother-in-law's cataract surgery serves to exemplify these problems. When she had her cataract removed, the procedure was performed under general anesthesia. Her consent form for surgery properly stated that, rarely, people die from anesthesia reactions. The dutiful admitting resident verbally presented the consent form's contents to her, and this fact was brought up by him as well. When my mother-in-law heard that she might die from the procedure, she refused to sign the consent, saying that she would rather have impaired vision than take a chance of losing her life. Both her family and the doctor responded by pointing out that although such disasters do happen, they are indeed very rare, and she should not overreact. We emphasized the fact that if an untoward event occurs even once in the history of a procedure, the patient must be advised of the possibility of its occurrence. We told her that this information was being presented more as a legal formality than a portrayal of a realistic possibility. After about a half-hour of persuasion, we eventually convinced her to sign the consent. She went to surgery the next morning and died of an anesthesia reaction.

Well, actually, she didn't die, and the surgery was successful and without complications. However, this incident, regardless of its outcome, exemplifies some of the inherent imperfections in the consent process.

In addition to these obvious imperfections, the process of consenting to procedures that, if the outcome is unexpected, can do serious emotional damage to the patient may be fundamentally flawed. In reproductive medicine there is no shortage of such situations. Consider the situation of surrogate motherhood. This treatment plan entails payment to a fertile woman for carrying the embryo of a woman who can produce eggs but who cannot carry the pregnancy herself. The costs of prenatal care for the surrogate are paid, and additional financial compensation is often included in such agreements. Of course, the signed agreement includes the provision that the surrogate will turn the baby over to the genetic mother after delivery. Under these circumstances, however, it is not unprecedented for the surrogate to decide once the baby is born that she is not emotionally capable of giving it to the genetic parent. The problem here is that the patient did not realize at the time the consent form was signed the extent of emotional trauma she would suffer when the time came to give the baby up. Although the notion that someone would renege on a signed contract can engender anger, it should be appreciated that the emotional stakes are extraordinarily high and it is understandable that people will not always be able to predict their own responses.

The relevance of this discussion to genetic manipulations that may have unexpected outcomes is that the consent process cannot provide complete protection for the patient or the doctor. If a patient agrees to a form of genetic intervention for any purpose, and the result is a devastating birth defect or neonatal death, the patient may well sue the caregivers, and juries are likely to be sympathetic to her grievances. If an adequate consent cannot be given and this is recognized by the caregiver, it is unethical to induce the patient to consent to the procedure.

Despite these concerns, much public discussion of the imminent use of genetic modification procedures has created the general feeling that many of the procedures are either ready for clinical use or soon will be ready. If we assume this to be true for the moment, we should evaluate these procedures from the perspective of the treatment candidate. How do genetic modification procedures measure up against alternative treatment options? Let's take a look at all of these options as they would apply to a typical clinical problem—avoidance of the genetic disease sickle cell anemia in the children of parents at risk.

The typical situation in which risk for sickle cell anemia in a child arises is where both parents are carriers of the disease. That is, they have one normal globin gene and one sickle gene, each inherited from a different one of their own parents. We will assume that the objective of such a couple is that most commonly adopted under such circumstances—the parents want to have a child who is not afflicted with the disease, and they both want to be the genetic parents of the child. We will consider three options for achieving these objectives. First, the couple can get pregnant, screen the fetus for the sickle cell disease genotype, and have an abortion if it is found that the fetus will develop the disease. Second, the couple can undergo in vitro fertilization (IVF), biopsy the embryos, perform PCR to determine the types of globin genes present in each embryo, and transfer only those embryos that will not develop the disease. Third, the couple can attempt some form of genetic manipulation. For the purposes of this discussion we will assume that the would-be parents are prepared to have a child who, like themselves, is a carrier of the disease. Although sickle cell disease is not strictly recessive, carriers have only very mild physiological changes. One of my college friends who starred on the track team was a carrier of the sickle trait, for example. Therefore, we will assume that the parents will be satisfied if the child either does not carry the trait or functions in a manner no worse than a carrier of the trait. We will also presume that the couple has no other significant health problems that would eliminate, for indirect reasons, any of our three treatment options. For example, an infertile couple could not choose option 1—conception, prenatal diagnosis and abortion—because they would be unable to conceive. We will thus assume that the health status of the couple allows them to freely choose any of the three treatment options.

In our evaluation of the three treatment options, we will use a modified "report card" format. We will examine three different phases of the treatment—establishing pregnancy, genetic testing to confirm success and rule out untoward events, and success in carrying pregnancy to term. Each of these three phases will be evaluated for safety, efficacy, morbidity, and cost, and an average score for each of these four parameters will be given. An average will also be given for the safety of all three phases of treatment, as well as for the efficacy, morbidity, and cost. These averages will yield a final numerical score that will represent the suitability of each approach for addressing the medical problem. If any aspect of treatment is considered outstanding from the point of view of patient well-being, that aspect of treatment will be given an "A," or a numerical 95. Any aspect of treatment that completely fails to meet patient needs will be scored with an F (55 or less), etc. There may be circumstances in which a grade of "F" for one part of the procedure renders the treatment option entirely unacceptable. For example, if an "F" in safety was given because of a high risk of death to a parent or to a newborn, the procedure would be regarded as ethically unacceptable and eliminated as a treatment option. When each of the various steps in the process of avoiding giving birth to a child affected with sickle cell disease is given a number, we will add up the numbers to see which procedure gets the best, or highest, score. With these rules established, let's now see how the various treatment options measure up.

If we look at the first option (Table 3), establishment of pregnancy and abortion if the fetus is determined to have the disease genotype (two copies of the sickle gene), then the procedures works out as follows. The couple seeks no medical intervention until after a pregnancy is established. The pregnant woman then submits to fetal testing by either amniocentesis or chorionic villus sampling (CVS), the latter being a procedure for sampling the fetal side of the placenta at an earlier time of gestation than is possible with amniocentesis. DNA is extracted from the tissue or cells retrieved and analyzed by PCR to determine whether the fetus is completely without the sickle mutation, is a carrier of the mutation, or is homozygous for the mutation and destined to develop sickle cell disease. The method used would be similar to that shown in Figure 17 (p. 56).

From the point of view of achieving pregnancy, this approach will receive an "A," or a 95.0, for safety, efficacy, morbidity, and cost. Getting pregnant by

Table 3   Report card 1: Ruling out sickle cell anemia by prenatal diagnosis and abortion

	Safety	Efficacy	Morbidity	Cost	Average
Achieving pregnancy	95 (A)	95 (A)	95 (A)	95 (A)	95 (A)
Genetic testing	90 (A-)	95 (A)	92 (A-)	85 (B)	91 (A-)
Pregnancy success	93 (A-)	75 (C)	84 (B)	94 (A)	87 (B)
Grade point average	93 (A)	88 (B+)	86 (B)	90 (A-)	**91 (A-)**

natural means is cost free and very effective, and, although being pregnant is associated with some morbidity (morning sickness, etc.) and risk (complications of pregnancy), these factors are minimally problematic when this technique is used. The genetic testing procedures, CVS or amniocentesis, are both very safe, although a small chance of pregnancy loss is a documented complication. We will give this component of the procedure a score of 90 for safety. For efficacy, this procedure receives a 95, as the success rate for both obtaining an adequate sample and performing PCR is very high. The procedure usually causes some anxiety and a small amount of discomfort, so we will give it a 92 for morbidity. The cost is about $2500, which, although low, is still substantial. Therefore, the cost component will receive an 85.

Concerning the next "pregnancy success" component of this option—preventing genetic disease by aborting the pregnancy if the fetus is destined to develop sickle cell disease—the evaluation is a little bit more complicated. It will be necessary to perform an abortion about 25% of the time, because this is the percentage of conceptuses that will have two sickle mutations. Abortions are very safe, and because they will only be done on 25% of the patients the safety component deserves a 93. Abortion is 100% effective, but if it is performed the overall effort to have a child free of sickle cell disease is a complete failure. Therefore, from the point of efficacy, we cannot apply a score of more than 75 (for the 75% of couples who will go on to have a disease-free child and who will not require an abortion). Although this rate of "failure" of the procedure is rather low by comparison with others, it is still quite significant. The morbidity from abortion varies because it is far more emotionally traumatic in some patients than for others. However, the emotional component must be viewed within the context of the fact that the couple would not choose this option if abortion was not acceptable to them. And, recalling the principle of medical practice that the attitudes of anyone other than the couple, including the care provider, are not a factor, we will judge the morbidity from abortion in terms of the physical suffering the woman must endure and the emotional suffering the couple will endure from having their effort to conceive a child free of sickle cell disease fail. However, these problems will affect only 25% of the couples, whereas the other 75% will enjoy an "A" for morbidity associated with the pregnancy. If we give a failing score of 50 to the 25% who must abort, then as a group this procedure will get a score of 84 or B for morbidity. The cost of abortion is quite low—a few hundred dollars—and, again, will be required in only a quarter of the cases. Therefore, we will give a score of 94 for the cost of disease prevention by abortion. The average score for the pregnancy success phase by this method is therefore 87. If we average all of these scores as shown in the report card, this approach to preventing sickle cell disease receives an A–, with an average numerical score of 91. In computing this average, the importance of safety, efficacy, morbidity, and cost were weighted equally.

Now consider the second option, IVF with embryo biopsy and a PCR test to determine the sickle genotype (Table 4). This procedure, which diagnoses genetic disease before implantation of the embryo, is termed preimplantation genetic diagnosis, or PGD for short. With PGD, the woman is given hormones to stimulate development of multiple eggs, induced to ovulate with HCG, and subjected to a procedure in which a needle is used to pierce the wall of the vagina and aspirate eggs out of ovarian follicles before their rupture. The needle aspiration is often performed under general anesthesia. The eggs are then fertilized, and when they reach 6–16 cells of development, they are biopsied for PCR testing. Surviving embryos that do not have two copies of the sickle gene are then transferred back to the woman in an effort to achieve pregnancy. If additional embryos of acceptable genotype are available after three embryos are transferred, these can be frozen and thawed later in another attempt to achieve pregnancy. These later attempts do not require another round of IVF; the frozen embryos are merely thawed and transferred to the woman according to a timing process that will allow the embryos to be at the blastocyst stage in the uterine cavity during the window of implantation of a spontaneous cycle. Use of frozen embryos is therefore far less invasive than the initial IVF cycle.

With PGD, the process of achieving pregnancy is clearly deficient in every respect to option 1. Hormonal stimulation has led to an occasional death, as has general anesthesia, which is often given for egg retrieval. The process of egg retrieval is also statistically safe but carries with it the risk that the sharp needle will cut into a blood vessel and cause bleeding. For safety we will accordingly give this option a B, or a numerical 85. What about efficacy? With this approach to ruling out genetic disease, pregnancy is by no means ensured. The worldwide data for the rate of establishing pregnancy after PGD show that pregnancy is successful in about 20% of cases. Thus, although failure will not

Table 4 Report card 2: Ruling out sickle cell anemia by in vitro fertilization with preimplantation genetic diagnosis.

	Safety	Efficacy	Morbidity	Cost	Average
Achieving pregnancy	85 (B)	40 (F)	80 (B-)	40 (F)	61 (D-)
Genetic testing	90 (A-)	95 (A)	92 (A-)	85 (B)	91 (A-)
Pregnancy success	94 (A)	95 (A)	95 (A)	90 (A-)	94 (A)
Grade point average	90 (A-)	76 (C)	89 (B+)	72 (C-)	**82 (B-)**

likely be due to the need to abort a pregnancy for avoidance of genetic disease, it is likely to occur as result of the fact that IVF does not always lead to pregnancy. Because successful pregnancy in option 1 occurs 75% of the time on average, we gave that procedure a maximum score of 75. Logically the, option 2 should not receive a score of more than 20, because successful pregnancy will not occur in more than 20% of cases. This "F" grade is probably lower than it should be, however. Worldwide statistics for PGD success include patients with infertility problems, because many seek this approach when infertility precludes selection of option 1. We may infer that fertile couples such as those in our hypothetical example will do a bit better with IVF because they don't have infertility problems that would interfere with IVF success for other reasons. Moreover, the ability to store some embryos for later transfer in the event of pregnancy failure can reduce the failure rate somewhat. With these factors considered, we will give this procedure a 40 for efficacy rather than a dismal 20.

Morbidity associated with establishing pregnancy is also greater than for option 1. The woman must inject herself daily with hormones for several days and undergo blood drawing to monitor her hormone response, possible ultrasound examination to evaluate development of her ovarian follicles, and egg retrieval. Egg retrieval involves needle aspiration of the ovaries under anesthesia, with a recovery period of several hours required for the procedure. This approach to establishing pregnancy is usually well tolerated but is clearly more traumatic than option 1, which requires only that the couple have intercourse to get pregnant. We will therefore give establishment of pregnancy by the PGD approach a score of 80 for morbidity. From the point of view of cost, option 2 does not compare favorably with option 1. IVF with PGD costs about $15,000 per cycle, about six times as much as pregnancy with subsequent prenatal diagnosis. We gave the $2500 option a score of 85 for cost, and so we will give IVF with PGD a score of 40.

The genetic testing procedure, which involves embryo biopsy and PCR, is completely safe for the woman and appears to be safe for the embryo as well. However, the PGD diagnostic procedure is a one-time test requiring successful PCR on the DNA from a single cell. For these reasons, the PCR test is always backed up by the standard tests used in option 1. Therefore, the safety of PGD genetic testing cannot exceed that of amniocentesis or CVS as performed in option 1, and it thus cannot receive a score higher than 90. Regarding the parameter of efficacy, the procedure is quite effective, with a worldwide failure rate of about 1%. However, the PGD testing procedure is relatively deficient in comparison to amniocentesis or CVS because PCR after embryo biopsy cannot be repeated in the event of technical problems and because the embryo biopsy allows testing only for the genetic disease and not for any other problems with the fetus. When amniocentesis or CVS is performed, sufficient material is obtained to allow dozens of repeat tests in the event that technical problems arise. Moreover, if the parents also wish the chromosomes examined for Down syndrome (three chromosomes 21 per cell) they have this opportunity. For efficacy of the test, PGD therefore receives a score of 84. However, we must remember that the PCR test is always backed up by standard testing as is used in option 1. Therefore, we will give the genetic testing phase of PGD a 95 for efficacy.

With regard to morbidity, the embryo test incurs none for the woman, and therefore the testing procedure receives a 95 for morbidity. However, again the PCR test must be backed up by standard testing, which received a morbidity score of 92 in option 1. Accordingly, this phase of PGD receives a 92 score as well. The cost of PGD is often factored into the total cost for IVF, so we will lower its mark for that component of the cost. But this procedure must be backed up by standard testing procedures, and we will therefore confer the same score, 85, for PGD testing as we give to the conventional approach of option 1. From the point of view of pregnancy success with avoidance of genetic disease, PGD is slightly safer than prenatal diagnosis and abortion because the minor hazards of abortion in the minority of patients who require it when option 1 is selected do not apply to PGD. However, IVF is usually accompanied by transfer of up to three embryos per cycle, which incurs a risk of multiple pregnancy. The complications of multiple pregnancy can be significant for both mother and fetus, and we will therefore give this component of option 2 a score of 94. PGD gets a high score of 95 with regard to achieving the desired pregnancy result. Although most patients will not get pregnant in a cycle of PGD, the low score associated with this reality has already been given at the step of achieving pregnancy. With regard to cost, we again give a good score to PGD, because the heavy costs of this process are incurred at the step of getting pregnant. Once pregnancy is achieved and selection of the correct embryos for transfer confirmed, a slight increase in cost might be incurred for managing the multiple pregnancies, because of a generally more vigilant management of these precious pregnancies. Accordingly, PGD gets a 90 for this component.

If these numbers are averaged, PGD receives a score of 82, or B–. Therefore, although this option is not a bad one, it clearly compares relatively unfavorably with the standard approach of prenatal diagnosis and abortion. PGD compares better with standard prenatal diagnosis if one parent has sickle cell disease and the other is a carrier of the disease. In this case, half of the conceptuses will have two sickle genes and would be aborted if the standard approach were taken. However, even in this case the standard approach scores better (a low B+) than PGD because of the enormous expense and morbidity of PGD and because of the low pregnancy rate associated with IVF and PGD.

Although it is not my intention to digress at this point and discuss issues of medical ethics, I feel compelled to point out here that I consider it the duty of care providers to fairly and objectively compare options 1 and 2 for patients and not to attempt to influence patients toward or away from the PGD option for any reason. Moreover, care providers should feel obligated to prevent their own personal views of abortion or embryo selection from influencing patients in any way. Patients must feel confident that their own personal decisions on these matters are completely acceptable to their care providers. The concerns of patients regarding the acceptability of such measures as abortion or embryo selection, the costs of medical care, and the sense of urgency that success be achieved as quickly as possible are the only factors that should influence the choice of treatment, and if care providers influence these choices through expressions, however subtle, of their own personal preferences, then the canons of medical ethics have been violated. I emphasize this point here because infertility specialists who perform PGD have an interest in performing this "high-tech" technique and may be predisposed to influence patients toward the choice of PGD. And of course, there are always care providers who oppose abortion.

The two approaches described above for ruling out sickle cell disease when parents are carriers of the trait are those presently used clinically. What about genetic manipulation of the germ line as a solution to the problem? Although these methods are not yet in use, the germ line gene transfer technologies described in Chapter 7 indicate that three major approaches of this kind might be possible: injection of genes into embryos, cloning of one of the parents, and establishment of embryonic stem cell lines that could subsequently be screened for their endowment of sickle globin genes, subjected to gene replacement if necessary, and then used for cloning.

Gene injection is a completely unsuitable approach to therapy in our hypothetical situation. When parents are both carriers of a genetic trait such as sickle globin where birth of a carrier is acceptable, only one of four of embryos conceived by such a couple would warrant genetic intervention. We certainly would not want to inject genetic material into an embryo that didn't need it, and thus use of gene injection would necessitate prior identification of those 25% of embryos actually in need of gene therapy. The only workable method for making such an identification at the present time is PGD by embryo biopsy. However, PGD requires that the embryo be at the 6- to 16-cell stage for biopsy and gene injection is only performed on the single-celled fertilized egg (see Figure 35 and p. 131 of text).[CE2] Therefore, it is impossible to know which embryos require treatment in time for employment of gene injection. We can therefore rule out this approach entirely.

What about cloning one of the parents? With this approach, the offspring would be genetically identical to one of the parents and would therefore be a carrier of sickle cell trait. This is an acceptable result, but there are other problems. This child would bear no genetic relationship to the parent who was not cloned, and therefore the objective that the child be genetically related to both parents would be defeated. From the point of view of the parent who was not cloned, the child might just as well be conceived with sperm donation (in the event the mother were cloned) or egg donation (if the father were cloned). Because donation of eggs or sperm is simple, relatively inexpensive (especially for sperm donation), safe, and effective, it makes no sense to clone one parent as an approach to ruling out sickle cell anemia in our example. Cloning might be used by producing embryos from the couple with the same methods as IVF, establishing ES cell lines from the embryos, studying the globin genes in the lines to identify those ES lines that were carriers of the sickle mutation or that had completely normal globin genes, and using the nuclei from those cell lines for cloning. Again, however, if one is to go to the lengths of producing embryos by hormonal ovulation induction and surgical egg retrieval, it is far easier to simply biopsy the embryos produced and perform PGD. Those embryos with the desired genes at the globin locus could be selectively transferred. Again, it would be poor medical practice to offer an untried and costly procedure such as cloning when an established procedure that works on the same principle is equally as effective already exists. Therefore, none of the existing genetic manipulation approaches has at least a theoretical chance of succeeding. To offer such approaches would at present be unethical medical practice.

To take a closer look at the various parameters of the genetic manipulation strategy, and to produce a report card for these procedures, let's consider a clinical setting in which the first two options, genetic diagnosis with abortion and PGD, cannot be effective. Such a situation would prevail if both parents actually had sickle cell anemia. In this case, the parents would both be homozygous for the mutation and would only produce sperm and eggs that carry the sickle globin gene. All offspring would therefore have sickle cell anemia, so no method of genetic screening, be it after implantation as in option 1 or before implantation as in option 2, can succeed. In this situation we can examine the possibility of microinjecting normal globin genes into embryos, or of establishing ES lines from embryos, performing a gene replacement to correct at least one of the sickle genes, and cloning. Let's first look at gene microinjection. The results are summarized in Table 5.

To obtain embryos for microinjection, ovulation induction and IVF must be performed. After microinjection of normal globin genes the embryos are returned to the woman in an effort to establish pregnancy. From the point of view of the safety the procedures for establishing pregnancy are therefore quite similar to PGD and, accordingly, receive a score of 85. With regard to efficacy of the step of achieving pregnancy, microinjection would probably be a bit less efficient than PGD. With PGD performed on two carriers of the sickle gene, 25% of embryos will not be transferred because they are homozygous for the sickle gene and the pregnancy rate for those transferred would be about 20%. With microinjection, some embryos will be destroyed by the mechanical trauma associated with microinjection and the remainder may be injured such that their chances of implanting are reduced. As we shall see shortly, many of the pregnancies established after gene microinjection will not be effectively treated for sickle cell disease by the gene transfer procedure. However, from the point of view of simply establishing a pregnancy, we can give such an approach a score of 30—slightly worse than for PGD. For similar reasons the cost and morbidity associated with microinjection are also similar to PGD (80 for morbidity, 40 for cost). Therefore, overall this process receives a score of 59.

The safety, efficacy, morbidity, and cost of the genetic testing procedure after microinjection are the same as for option 1—getting pregnant and performing amniocentesis or CVS. In the case of microinjection the test methods are the same, but the test is for integration of the injected globin gene. This test is straightforward, reliable, and relatively inexpensive. Overall, the testing phase receives a score of 91.

With regard to pregnancy success, or achieving the desired result from the pregnancy, the safety is probably a little bit higher for microinjection than PGD or even option 1, standard prenatal diagnosis. This is because nearly all of the pregnancies will require termination because of failure of gene integration or in-

Table 5   Report card 3: Curing sickle cell anemia in embryos of parents with disease by gene microinjection.

	Safety	Efficacy	Morbidity	Cost	Average
Achieving pregnancy	85 (B)	30 (F)	80 (B-)	40 (F)	59 (F)
Genetic testing	90 (A-)	95 (A)	92 (A-)	85 (B)	91 (A-)
Pregnancy success	94 (A)	5 (F)*	55(F)	93 (A)	41 (F)
Grade point average	61 (D-)*	43 (F)*	89 (B+)	75 (C)	67 (D+)*

tegration of inappropriate numbers of copies of the new globin gene. These issues are elaborated on below in the discussion of efficacy, but suffice it to say here that abortion is safer than pregnancy and the large majority of these fetuses will require abortion. Therefore, gene microinjection is the safest option thus far.

What about the efficacy of gene injection? Recalling our results from animal experiments, we expect that at most 15% of the injected embryos will actually sustain integration of the injected DNA, whereas in the remaining 85% microinjection will fail. Moreover, as pointed out in the discussion of gene injection (see pages 135–136 to review this), it is very unlikely that those embryos that sustain integration will express the new gene at the appropriate level. More likely than not, many copies of the new gene will be inserted and expression will not be at a perfect level for therapy of the disease. Moreover, there are always some integration events that are associated with complete expression failure. Another serious problem is the potential for foreign DNA integration to disrupt host gene sequences or even break chromosomes. As pointed out above, these phenomena have been well documented in animals and associated with serious developmental problems and even death. A baby born with any of these problems would be considered in the group of "failures" with regard to "efficacy" as defined by obtaining the desired pregnancy result. Because efficacy is so low, and because the manifestations of failure could well be intolerable developmental disasters, the efficacy parameter has been given a score of 5 and marked with an asterisk to indicate that the consequences of failure are completely intolerable. The presence of the asterisk thus renders this method totally unacceptable, because there is no consent process that could adequately protect the patients from the adverse consequences of the procedure. The patients may say they understand and accept these risks, but more likely than not those who sign such a consent form are not fully aware of the impact that adverse consequences are likely to have on both the health of their baby and their own emotional state. Accordingly, microinjection must be considered a medically unethical and unacceptable approach to the problem. The final score is also marked with an asterisk to reinforce this point.

Although the single problem of the lack of efficacy of gene microinjection renders the procedure unacceptable as a treatment choice, a few other features of this strategy are worth mentioning. Regarding morbidity scoring in the area of pregnancy success, it is important to note that the vast majority of these patients are likely to require abortion. In option 1 we scored this parameter as a 70 on the assumption that abortion would be quite traumatic for the 25% of patients who would require it. In the case of gene microinjection the percentage needing abortion would be at least 85%, because 85% of embryos would not retain the injected gene, and probably over 90%, because many conceptuses that retained the gene would have other problems such as inappropriate gene copy numbers. Therefore, morbidity here is scored as 55, or F. The cost of bringing a successful pregnancy to term with this procedure might actually be a bit better than with PGD, because most of these pregnancies will be terminated after genetic testing. Because term pregnancy is more expensive than abortion, we can predict that gene microinjection, on average, would cost less than PGD. However, the cost might be slightly more than for prenatal diagnosis and abortion because of the special monitoring that these pregnancies would require. Overall, the "pregnancy success" component of this procedure would receive a failing score of 41, and the procedure as a whole would be unacceptable.

If we can't microinject, can we clone? A cloning procedure would allow both parents to contribute genetically to the child if embryos were produced and used to make cell lines. Moreover, a gene replacement could be performed in the cells, thus producing a disease-free child who was indistinguishable at the molecular level from a typical carrier of the trait. Before we produce a report card for cloning, an overview of the logistical challenges presented by this approach would be useful.

The data from animal cloning experiments indicates that about 1 in 50 embryos produced after removal of the egg chromosomes and transfer of a nucleus actually develops to the point of being born alive. However, only about one in five nuclear transfer procedures results in successful creation of an embryo with implantation potential (as a reminder, these numbers are shown in Table 2, on p. 147). If we are to clone a human, then, we must realistically plan to obtain 250 eggs for manipulation. Fifty of these eggs will survive manipulation to become embryos, and one of those embryos, once transferred, will lead to a live birth.

Where are we going to get 250 eggs? Obviously we will need egg donors. Egg or ovum donation, as it is referred to in the assisted reproduction field, is not unusual. Quite often, women are unable to produce their own healthy eggs, so female egg donors are paid to undergo hormonal ovulation and surgical egg retrieval. These women may be paid as much as $10,000 to endure the risks and morbidity associated with ovum donation. Although the number of eggs a woman will produce in ovum donation varies with the individual, we may optimistically predict that on average we can get 10 good eggs from each woman. Under this assumption, cloning will require recruitment and payment of 24 female egg donors. Number 25 can be the woman who actually wants to have an embryo cell line cloned.

Once the 50 healthy embryos are produced from the 250 donor eggs, it will be necessary to transfer them back to the uterus of a woman in a timed manner, such that the embryos reach the blastocyst stage within the window of implantation. In human assisted reproduction, it is unwise to transfer more than three embryos into any one woman. If more than three embryos are transferred and they all implant, a potentially disastrous multiple pregnancy could result. In the setting of cloning we know from data with animals that very few of the 50 embryos will develop to an advanced stage, but we can't afford to take the chance that multiple pregnancy will threaten the life of the fetus that we hope will develop to term, and we certainly do not want to endanger the life of the woman, either.

If we're going to transfer 50 embryos to the uterus in groups of 3, we can take either of two approaches. If the woman who wants the baby is the only person to receive transferred embryos, then we can freeze 47 embryos and transfer 3. If the first three fail to develop we can return to the freezer, thaw three more, and perform another transfer. This strategy is workable but very cumbersome. First of all, although embryo freezing works quite well, it does not work perfectly, and some of the 50 embryos are certain to be damaged in a freeze-thaw. Moreover, if the woman gets pregnant but loses the pregnancy, as occurs frequently when animals are cloned (see Table 2 on p. 147 again), she will have to recover for some period before undertaking another embryo transfer. It should be readily apparent from these logistical realities that having a baby by this method could take years, and the chances of success would be lessened a bit by the freezing step.

Another approach would be to recruit women to serve as surrogate mothers, carrying the embryos for a fee. Surrogate motherhood is not unprecedented. Typically, surrogate mothers are paid a fee, which we will estimate in present-day dollars to be $20,000. For surrogates to carry the reconstructed embryos, but not carry more than three of those embryos per woman, it will be necessary to recruit 16 women as surrogates. Once the surrogates are recruited, another problem arises, however: How are we going to make sure that the timing of embryo transfer is appropriate from the perspective of "hitting" the window of implantation in every surrogate? Clearly, synchronizing the cycles of 16 women such that they will all be within the window of implantation within the same several-day span is a severe logistical challenge. Although it is not impossible to achieve this goal, we must realistically accept how difficult it would be and plan to freeze at least some of our 50 embryos for a short time. If freezing is going to destroy some embryos, we probably should go back to the drawing board and plan to obtain a few extra eggs for creating embryos by nuclear transfer. Recalling that we need 24 egg donors in addition to our patient to obtain 250 eggs for manipulation, it is perhaps wise that we recruit 26 or 27 women instead. With 280 eggs for manipulation we can plan on getting 56 embryos and hope that no more than 6 die in the freeze-thaw process.

It should be apparent from this discussion that cloning in humans is logistically exceedingly complex. Why is it so easy in animals and so tough in humans? First, it is not so easy in animals. As examination of the tabulated data (Table 2) in Chapter 7 shows, the success rate in animals is exceedingly low. However, some problems with cloning in humans do not apply to animals. In many of the species thus far cloned, like pigs and mice, it is normal for the female to produce large litters. Because of this, transfer of large numbers of embryos produced by nuclear transfer is not inconsistent with the normal physiology of the animal. Another important mitigating factor in animal cloning is that the fear of multiple pregnancy is not as great as it is for humans from the perspective that morbidity or mortality of the female is not as disastrous as it would be in humans. In the pig cloning study referred to above, some of the females received more than 100 embryos in a single transfer! This was done on the assumption that few of those embryos would actually implant. The ability to insert so many embryos into a single animal alleviates the problem of finding multiple carrier females as would be necessary for humans. Thus, for a variety of reasons, cloning animals is far simpler than cloning people.

With this background, let's now put together a report card for cloning as a treatment for sickle cell disease. This "report card" is shown in Table 6. To briefly review the strategy, our plan would be to produce embryos from two parents with the disease and establish cell lines from these embryos. A targeted gene transfer procedure would then be conducted that resulted in substitution of a normal globin gene for one of the sickle genes. The nuclei of these genetically modified cells would then be used to replace the chromosomes of unfertilized eggs to create embryos. The embryos would then be transferred to women in an attempt to obtain a newborn who would be the genetic child of both original parents but who would not have sickle cell disease.

Table 6   Report card 4: Curing sickle cell anemia in embryos of parents with disease by gene establishing cell lines from embryos, performing gene replacement, and cloning.

	Safety	Efficacy	Morbidity	Cost	Average
Achieving pregnancy	60 (D-)	5 (F)*	60 (D-)	10 (F)	34 (F)*
Genetic testing	98 (A+)	99 (A+)	100 (A+)	50 (F)	87 (B+)
Pregnancy success	40 (F)*	10 (F)*	10 (F)*	10 (F)	18 (F)*
Grade point average	66 (D)*	38 (F)*	57 (F)*	23 (F)	46 (F+)*

First, we will examine the step of achieving pregnancy. From the safety standpoint, cloning is not very good. The would-be mother must first undergo ovulation induction and IVF for production of the embryos that would be used to establish the cell lines. This process is as safe for her as PGD was for the parents who carried the sickle trait. However, in this case, the woman would have to undergo another round of ovulation induction when the genetic manipulation of one of the cell lines was completed and the time arrived to perform nuclear transfer. This is because she is serving as one of the egg donors. The genetic manipulation phase would require a few weeks, so it would not be possible for the woman to produce eggs for both establishment of cell lines and nuclear transfer in a single cycle of ovulation induction. Thus the risks associated with ovulation induction and egg retrieval would be doubled. This is not the only parameter of safety in the case of cloning, however. Let us not forget the other 25 or so women who will be required as egg donors for this procedure. All of them would be subjected to the risks of ovulation induction and egg retrieval, and so the risks of a PGD cycle (as a reminder, option 2 above) would actually be multiplied by 25 or so for cloning. For these reasons we will give the safety component of achieving pregnancy a score of 60.

With regard to the efficacy of achieving a viable pregnancy, we need only be reminded that only 1 of 50 of the transferred embryos will yield a pregnancy with the potential for full-term development. If we give PGD a score of 40 for efficacy because the pregnancy rate per embryo transfer is only about 20%, what score shall we give to cloning, where the rate is about 5%? (Incidentally, we obtain the figure of 5% from the fact that about 20 embryo transfers of 3 embryos per transfer will be performed to obtain a viable pregnancy). A reasonable score would perhaps be 10. However, at this point we must consider the fact that these viable pregnancies are at extremely high risk for disastrous developmental anomalies that could lead to early neonatal death or severe birth defects. Recall that in cattle fatal birth defects occurred in 25% of offspring and in mice, 10 of 17 newborns died within a week or so of birth. Moreover, because cloned mice that appear grossly normal in fact have very disordered gene regulation and a reduced life span, a high probability exists that none of the human offspring would be entirely normal. For these reasons we will give the efficacy component of achieving a pregnancy a score of 5, but we will mark it with an asterisk to indicate that problems associated with this step render it clinically unacceptable.

From the point of view of morbidity, achieving pregnancy by this cloning method is again similar to PGD, but the most unpleasant parts of the process must be endured by all egg donors. So we will assign a score of 60 for morbidity. Regarding costs, we may estimate a cost of at least $200,000 if all egg donors are paid several thousand dollars apiece. Accordingly, we will give a score of 10 for cost.

Regarding the genetic testing phase of gene replacement and cloning, we have a situation in which cell lines can be maintained for whatever period of time is needed to obtain a reliable result and the result would be unequivocal. Because the genetic tests would be performed on cell lines they would be completely safe, and the tests are so definitive that it even might be possible to skip the confirmatory fetal genetic testing. So we will give the safety component of genetic testing a 98. By similar reasoning, the efficacy of the genetic testing will receive a 99. False results on tests for targeted integration are very rare indeed. Genetic testing on cell lines results in no absolutely morbidity for the would-be parents, so we will give a score of 100 for this component of the genetic screening procedure. The cost of testing for targeted integration is high. Recall from our discussion of this methodology (see pp. 126–128), that on average, only about one in one million cells exposed to foreign DNA sustains a targeted integration event. To test hundreds of clones for targeted integration by PCR, to confirm the findings, and to ensure that the chromosomes in the cell lines have remained normal after long-term culture, we will be required to spend many thousands of dollars. We will therefore assign a score of 50 for cost.

From the point of view of pregnancy success, we must provide a failing score, and again we must mark the safety component with the "asterisk of unacceptability." Failure of full-term development is often safe because an advanced pregnancy is never achieved. However, as we have seen from our review of cloning experiments in animals, pregnancy losses often occur at a late stage. Late pregnancy loss in humans can be very dangerous for the woman, and so we will assign a score of 40 (some pregnancies will be lost at an early stage) and include the asterisk. For efficacy, we are reminded that only about 1 of 50 transferred embryos will develop to term, a rate of 2%. We will therefore give a score of 10 to this component. If we recall our background discussion of the importance of efficacy in determining the acceptability of a medical procedure, we will mark this component with an asterisk to indicate that efficacy is so low as to render cloning unacceptable as a choice.

Regarding morbidity associated with carrying a pregnancy to term, there will be no morbidity if the baby is normal but physical and emotional morbidity will be substantial from late pregnancy losses or neonatal deaths. We will accordingly give a score of 10 for morbidity and mark it with an asterisk as well. The degree of suffering likely to be endured if a newborn dies in its mother's arms is unacceptable. And, finally, where cost is concerned, we must anticipate that the 16 surrogate mothers will each require a $10,000 to $20,000 fee for undergoing the embryo transfer. Note here that this cost is separate from, and additional to, fees paid to egg donors. Accordingly, we will give this component a score of 10, but we will not indicate the cost to be completely unacceptable, because there will be some people who are willing and able to endure the cost. Overall we can see that, although cloning and gene replacement can work in principle, in practice it cannot be justified or carried off logistically.

It is important at this point to emphasize some of the deficiencies of the existing cloned animals as models for human cloning. A mouse with many abnormally regulated genes may look perfectly normal, and most cloned mice do in fact appear normal. However, if the same genetic anomalies present in these animals existed in a human being, they could manifest as profound developmental abnormalities. Consider, for example, that we cannot use mice or the other animals cloned thus far as models for psychosis or other severe mental disorders, nor we can we use them to model other serious problems such as autism or mental disability. Because a substantial number of human genes contribute to development of higher brain functions that have no counterpart in these animals, the identification of gene regulatory defects in cloned animals shows unequivocally that the procedure cannot be used in humans. Thus, from the point of view of efficacy, we must give a score of 5 and mark this component with an asterisk to indicate that the procedure is inconsistent with the canons of ethical medical practice (see report card 4). I emphasize here that the philosophical question of whether reproductive cloning is "moral" has not been addressed. I have merely pointed out that, given the existing preclinical data from animals, the use of reproductive cloning as a medical procedure would be malpractice.

Another way to appreciate that cloning would be clinically unacceptable is to examine the procedure from the point of view of the informed consent process. Any such consent form would have to advise the women who carried the embryos that they might have as much as a tenfold increased chance of losing a pregnancy relatively late in gestation (remember the sheep?). Although these women might be willing to sign a form acknowledging this risk, their signatures would not carry much weight with a jury in a malpractice case if a surrogate sought reparations for the severe emotional stress or injury that might accompany late pregnancy loss. Remember that these surrogate embryo carriers are only performing a service, but that service is one that puts their own physical and emotional health at significant risk. The surrogate carriers as well as the would-be parents would also have to be advised that, if born, the baby could have as much as a 40% chance of dying shortly thereafter (remember the mice?). The emotional stress of late pregnancy loss is severe but is far less traumatic than the death of a newborn. One might ask at this point why the surrogate women would have to be informed of this risk if their role is to simply hand the baby over to the parents after their service of carrying the pregnancy is completed. In contested surrogate mother cases, in which the woman who carried the pregnancy subsequently sought custody of the child, her emotional bonding with the baby has been regarded as significant by the courts and has mitigated against the legal weight of her signature on a contract that gives custody of the baby to the genetic parents. If these court rulings are to be taken seriously, then the surrogate mothers in a cloning procedure who might actually be delivering the child must be informed of the emotional risk of neonatal death. Again, there is no consent form acknowledging such a risk that would hold up in court if the aggrieved woman, whether she be the biological mother or a surrogate, filed suit against the medical team.

Although a more extensive discussion of the ethical and legal aspects of germ line genetic modification will be undertaken later, a couple of points should be clear from the present analysis. First, although it is interesting and provocative to discuss the "deep" philosophical issues surrounding engineered changes in the human genome, such discussions should not be undertaken with the purpose of deciding whether it is ethical or acceptable to perform genetic engineering. At present, there exists no genetic engineering methodology that would be ethically acceptable as a medical procedure, for far simpler and more practical reasons than those that would be the subject of a philosophical debate. Genetic engineering is unethical medical practice, and it is therefore presently irrelevant whether it would violate some deeper moral tenets.

Another point to be appreciated is that because genetic modification methods would be medical malpractice, it would be impossible in the United States for practitioners to obtain malpractice insurance coverage for these procedures. Thus there are natural controls on technologies such as cloning that preclude their use within the United States. It is therefore not necessary to make such procedures illegal, because they cannot in any case be carried out. Of course, an irresponsible group of physicians and scientists might violate accepted standards of medical care and attempt genetic modification on a patient or patients, but these practitioners would already be practicing medicine without insurance. Moreover, persons willing to act so irresponsibly would not likely be deterred by additional legislation prohibiting the use of cloning or any other form of deliberate genetic modification. As with the ethical issues surrounding germ line genetic modification, the legal issues will also be addressed more extensively later. Suffice it to say that mechanisms are already in place to prevent germ line genetic modification, in its present state of refinement, from being used in the United States. And, of course, no government can exert jurisdiction over activities that take place outside its borders.

Although we can dismiss current germ line genetic manipulation procedures as medically unethical and/or logistically impractical and, in doing so, avoid the larger questions about whether such procedures are socially acceptable, to do so would be unsatisfying for two major reasons. First, it can always be argued that although genetic manipulation technology is currently primitive there will certainly come a time, perhaps in the not so distant future, when it becomes efficient, safe, and practical. Because this happenstance would then force a discussion of the larger issues, why not have the discussion now? Wouldn't a proactive airing of the issues make development of a coherent and reasonable policy on genetic engineering more attainable? Another simple fact is that it's just plain fun to talk about futuristic scenarios in which parents shop for the genotype of their child. A discussion of genetic engineering that ended perfunctorily with the conclusion that it will never happen would deny us all the opportunity to engage in the intriguing, fascinating, even awe-inspiring speculation about a world in which one can not only obtain designer jeans, but designer genes as well!

To set the stage for the broader ethical discussion that is relevant to future, more refined genetic engineering technologies, we must have some idea of what will and will likely not be possible when such technologies are in place. The discussion thus far has been predicated on the assumption that just about anything is possible. However, there may well be theoretical and practical limitations imposed by the biology of human development that make some scenarios very unlikely. In the next chapter some of these potential limitations will be reviewed. Once they are understood, a more clear and meaningful framework for an ethical discussion will have been put in place. Another important aspect of futuristic discussions of human genetic engineering is that these procedures will always be, at their core, medical interventions. As such, it is meaningful to discuss them only in the context of clinical decision making, as we have done for our present-day case of sickle cell anemia. That is, genetic engineering may well be one of several treatment options available and will have to be measured against those other options for safety, efficacy, morbidity, and cost.

With these ground rules in place, let's take a look at the future.

9

FUTURE DEVELOPMENTS AND APPLICATIONS OF GENETIC ENGINEERING TECHNOLOGY

Inevitably there is a gap in time, sometimes great, sometimes minute, but always measurable, between the occurrence of events and our perception of them. Therefore, we live our lives entirely in the past. The future, even the present, are mere products of our imagination.

—Me, just now (or rather, just a short time ago)

When I was in grade school, in the late 1950s, our public school received a visit from a spokesman for General Motors. The purpose of his visit, which was highlighted by a talk to a large student assembly, was to tell us youngsters what the experts at his company were predicting for the future of automobile travel. His speech, embellished by a very engrossing slide show, described the automobile of 1990.

The cars in his slides looked like small spaceships. They were very sleek in design, and they didn't have wheels. Instead, they floated on a cushion of air. By 1990, it was asserted, the need for tires, or even roads, would long since have passed. Nowhere in the talk was there any mention of fuel shortages, emissions standards, air bags, or global warming, and nowhere in his talk was there a description of how technological developments at the time led him and his colleagues to make their predictions for auto travel 30 years later. However, nobody doubted that his predictions would come to pass. Hadn't cars gone from lumbering boxes with running boards to lavishly finned vehicles with double headlights in just a few short years? Hadn't the engine performance of the typical passenger car improved fantastically over those same few years? Given the progress at the time, why shouldn't we have assumed we would be floating on air three decades hence?

When I first saw *2001, A Space Odyssey* in 1967, I took this spectacular visual experience to indicate what we should expect from space travel in the year 2001. We would, of course, have a permanent base on the moon that functioned much like a small city, spaceships that traveled to Jupiter while their passengers hibernated, and computers with personalities. Who would doubt, with the first moon landing on the immediate horizon, the practicability of the technologies depicted in this wonderful film?

The purpose of these vignettes, of course, is to illustrate how difficult it is to predict the technology of the future on the basis of developments in the recent past and further, how the tendency is to overestimate what progress will be. Why are these errors so easily made? First, technological "breakthroughs" often appear to come out of nowhere when in fact, they are usually manifestations of slow, steady progress. When we first introduced genes into the mouse germ line in 1980, much of the scientific community, as well as the popular press, were thunderstruck. Major articles announcing this development as a gigantic leap forward appeared on the front pages of leading newspapers and nationally circulated news magazines. Suddenly, as if out of nowhere, we stood at the threshold of a new era in genetics.

Although the development of transgenic mice certainly did provide new opportunities for both basic and applied experiments in genetics, it did not come out of nowhere as it appeared to do. Advances in gene cloning, as well as successful transfer of genetic material into tissue culture cells, in fact paved the way for transgenic animal technology. Many of these road-paving experiments went virtually unnoticed by the public, which is entirely understandable. As a result, the transfer of genes into an animal seemed like a huge leap forward. Some prominent scientists have in fact suggested that science moves forward in "spurts, " with intervening periods where novel findings are few, and where "fill-in" experiments are done to fully understand and characterize the recent breakthroughs. This illusion often leads one to think that major technical obstacles in front of us today will be solved by sudden forward leaps. Because these forward jumps are believed to come out of nowhere, it is only natural that we are unable to envision exactly how serious problems will be solved. We are satisfied, subconsciously if not openly, that they will be solved, somehow, by a breakthrough.

Another reason that predicting the future is difficult is that social developments that can dramatically affect the rate and direction of scientific progress are very difficult to predict. As alluded to above, issues such as global warming, fuel shortages, and passenger safety were not considered significant in the 1950s, but they have a major influence on automobile design and production today. Similarly, health researchers had no idea in 1975 that in 1985, a vast amount of their talent would be diverted from problems like cancer to the AIDS epidemic. Just two years ago anthrax was just an interesting disease. Now it and related potential agents of biological warfare are the objects of far more intensive investigation.

It is in the context of these realities that we should consider the future of genetic engineering technology. We've all heard it said that we will soon be determining the genetic makeup of our offspring either through gene transfer or genetic selection, as portrayed in the recent film *Gatacca* (1997). Many respected scientists are whispering that the process of human evolution will soon be under our control and will no longer be driven by the forces of natural selection. Are these predictions realistic, or are they in fact as far off as was the space travel technology depicted in *2001, A Space Odyssey* in the year 1967?

The answer, of course, cannot be known for certain. The future is, after all, an illusory construct of minds that spend all of their time in the past. But we can attempt to arrive at realistic predictions by taking some simple precautions and by reminding ourselves of some well-known facts about human development. The most important precautionary measure is to refuse to assume that obstacles that seem insurmountable today will simply be circumvented by some major discovery. If there are problems standing in the way of developing acceptable genetic engineering approaches that are now without a clear solution, we will assume these problems will still beset the genetic engineers of the future. Another important precaution is not to forget that genetic engineering will necessarily be implemented as a medical intervention and, as such, will inevitably be measured against other invasive approaches for addressing the concerns of the candidates for therapy. These alternate medical approaches will also be improved with the passage of time, and they may well remain the most favorable options for controlling the genetics of our children.

The Future of Gene Transfer Technology

What improvements must be made in gene transfer technology to make germ line gene alteration into a clinically acceptable procedure? What's on the drawing board for improving gene insertion and gene replacement methods, and which of these improvement strategies is likely to pan out and develop into an efficient and safe approach to germ line genetic modification?

If we recall the discussion of direct embryo gene transfer methods—microinjection of genes into the pronucleus of the fertilized egg and use of recombinant viruses for infection—we have the following current deficiencies: These procedures are inefficient, achieving successful gene insertion in fewer than 20% of manipulated embryos; they lead to random integration of genetic material into the chromosomal DNA, which could, of course, be associated with disease-causing insertional mutations in host genes or host gene regulatory sequences (that is, a foreign gene insertion could separate a host gene from its promoter or destroy one of its key enhancers). Finally, in the case of gene microinjection, we have no control over the number of copies of the new gene that will integrate into the host genome. It is not unusual for a transgenic mouse to sustain insertion of several hundred copies of the new gene, linked together in head-to-tail fashion. Such insertions are not likely to produce the desired level of foreign gene expression.

Where gene microinjection is concerned, there are few new findings that suggest a solution to the problems of low efficiency, random integration, and lack of control of gene copy number. In keeping with our commitment not to simply take it on faith that these problems will somehow be solved, we will therefore conclude that human germ line gene transfer, if it is ever done, will not be done by this method. Where recombinant viruses are concerned, the situation is a little bit more promising. Many viruses have evolved to replicate their genetic material by minimally disruptive insertion of a single viral DNA molecule into the chromosome. There is also some evidence that viruses do not integrate randomly. In the case of retroviruses, the class of viruses including the AIDS-causing virus, HIV, integration appears to occur preferentially into DNA that is within, or very close to, an active gene. That is, the sites of retroviral insertions, when studied in early embryos (the time when the viruses are inserted), are relatively free of chromosomal proteins that tend to shield DNA from enzymes that activate genes and decode them into messenger RNA. This feature would of course be highly undesirable if we wished not to disrupt the function of host genes. However, the fact that these viruses do not integrate randomly suggests that mechanisms for nonrandom insertion might be modified such that the viruses can be directed to integrate into sites far away from any active gene. It is important to note in this regard that the human genome has far more DNA than is needed to house the 40,000 or so genes and their associated regulatory sequences. In effect, the human genome can be regarded as a DNA "ocean" with functional genes scattered within it like buoys. Therefore, it is presumed that many sites exist in the genome where insertion of new genetic material would have no effect on the function of any host gene.

Another interesting case is that of the adeno-associated virus, abbreviated AAV. This is a very small DNA virus that causes no known disease in humans and cannot replicate without "help" from another virus that coinfects the cell. Under some conditions AAV has been found to integrate into a specific region of human chromosome 19. Although the preferred AAV integration site is actually close to a gene expressed in muscle, this mechanism of nonrandom integration appears entirely different from that of retroviruses and again suggests that viral integration mechanisms, if better understood and exploited, might be modified so as to target recombinant viruses to designated benign sites in the genome. No obvious approach to such exploitation yet exists, but at least we have some evidence from the behavior of these viruses to speculate that such improvements are at least possible.

There is also evidence that gene transfer using recombinant viruses might be developed so as to improve efficiency. At present, the production of transgenic mice with recombinant viruses is about as efficient as with microinjection. This success rate of gene transfer in about 10% of exposed embryos is too low for clinical use, but new methods are being developed to provide viruses with novel coat proteins that could eventually increase dramatically the efficiency with which these viruses interact with the surface of the early embryo. If such improvements could be made, such modified, or "pseudotyped," viruses could insert far more efficiently than the present-day viral gene transfer vectors. Thus recombinant virus technology offers some promise as a possible future germ line gene insertion approach, although much work remains before it reaches the point of clinical adequacy.

In our discussion of gene transfer in embryonic stem cells above, we pointed out that a perfect gene replacement occurs, albeit rarely, in cells exposed to purified, or "naked," DNA that has a gene sequence very similar to a corresponding chromosomal sequence. Is there anything on the drawing board for extending this methodology directly to embryos? If this could be done, embryos could be simply exposed to DNA for correction of genetic abnormalities. If the methodology were efficient and safe enough it would not even be necessary to test the embryos for genetic disease. Instead, all embryos could be subjected to gene transfer, and those that did not require it would simply substitute the normal transferred gene for their own normal gene and would thus be genetically unchanged. Unfortunately, there is no technology even on the horizon that would increase the gene transfer frequency (presently successful in about 1/100 exposed cells in the best cases) to acceptable levels, and there is no method on the horizon for eliminating the 999/1000 of those 1/100 gene transfer events that are not perfect substitutions of donor for recipient cell DNA. Recall from the discussion of gene replacement in ES cells that methods were developed for sensitively detecting the 1/1000 perfect substitution events, not for eliminating the unwanted ones.

In keeping with our commitment not to assume that major obstacles will be overcome when there exists no evidence to suggest even a strategy for overcoming them, we will assume that, even in the future, gene replacement procedures will require the tissue culture environment. If this is true, then genetic changes introduced by this method will require major improvements in cloning. The technology exists today for performing gene replacement in tissue culture, but the use of the genetically manipulated cells for nuclear transfer into eggs is, as pointed out in the previous chapter, woefully inadequate for human use. It is reasonable to speculate, however, that the efficiency and safety of cloning can be improved. Experimental strategies do exist for identifying embryos that will in fact develop normally after nuclear transfer, and research with tissue culture techniques indicates that it may be possible to modify the tissue culture conditions in such a way as to make the cells more suitable as nuclear donors. By "suitable," I mean capable of reprogramming their genes after nuclear transfer quickly and completely enough to normal early embryo development and to produce offspring with normal gene regulation.

Although such advances in cloning technology are many years, perhaps decades, away, it might be possible in the future to substitute one or even many genes in cell lines established from embryos, after which the nuclei of those cell lines could be used for a cloning procedure. A little bit later we will discuss how the development of this capability might actually be applied to the genetic manipulation of the human germ line. Suffice it to say for now that there is no apparent strategy for performing multiple gene substitutions in a single gene transfer event. If we want to change many different genes in a cell line established either from early embryos or from fetal or adult tissue, these changes will have to be performed one at a time. It should also be appreciated that even if cloning is dramatically improved it will always require retrieval of eggs for nuclear transfer as well as costly and technically challenging manipulation of reproductive cells under the microscope. As such, cloning will always be relatively costly and invasive.

There are some other new methods of gene transfer on the horizon that might be applicable to germ line genetic manipulation. The most promising of these entails construction of an artificial chromosome that carries only the genes one wishes to transfer into the cell and then insertion of the entire chromosome. These artificial chromosomes have their own centromeres, which are required for their proper segregation during cell division (see p. 67). If one could put a whole new chromosome into the embryo, the risk that the transferred genetic material would physically disrupt important host genes on other chromosomes would be avoided. Moreover, strategies have been suggested for engineering the new chromosomes in such a way as to induce their loss from cells when they are no longer wanted. From the point of view of safety and proper regulation of the donor genetic material, this strategy looks attractive.

It has been possible to obtain stable maintenance of artificial chromosomes in tissue culture for many cell divisions. However, stable maintenance throughout the development of a human embryo is a distant goal that may never be achieved. Moreover, we do not know as yet whether a new chromosome could break and rejoin with others of the host chromosomes, a process known as translocation. If a piece of the donor chromosome became translocated to one of the host chromosomes, the results could be disastrous. Therefore, much more research is needed before this technology becomes reliable and safe. It may in fact never be possible to introduce artificial chromosomes into embryo and have them persist stably and safely for the millions of cell divisions the embryo will undergo, and for the many years that cells will function in the adult.

Another recent advance in germ line gene transfer has been the successful introduction of new genetic material into developing sperm. This procedure, which has been used thus far in the mouse, entails removal of early sperm-producing cells, exposing them to DNA, and then transplanting them back into the testicle of a recipient mouse. After transplantation the genetically manipulated cells resume sperm production and eventually give rise to functional sperm that carry the new genes. From the perspective of human germ line gene transfer this technique is very primitive in a number of aspects. It is not very efficient, and it has worked only with modified retroviruses. After the spermatogonia, as the early spermatogenic cells are called, are returned to the testis, the mature sperm they produce are in such low numbers that they have no chance to fertilize eggs unless the other, unmanipulated sperm that are produced in the testis are dramatically reduced in number. The testis that receives the transplanted spermatogonia can in fact be depleted of its own sperm either by exploiting mutations that block spermatogenesis or by giving the animal drugs that destroy its sperm. Procedures such as these are difficult to extend to a human transplant recipient. However, it is not impossible that in the future, spermatogenic cells will be cultured for sufficient periods to allow gene transfer and even gene replacement using naked DNA and that methods will be devised to produce mature sperm either in tissue culture or in an animal. Although sperm derived from the genetically manipulated cells would be few in number, the advent of ICSI, the procedure whereby individual sperm are injected into eggs to achieve fertilization, allows conception to occur when very few sperm are available. Thus we might envision a time when spermatogenic cells are targeted for germ line gene manipulation. This technology would not allow alteration of genes inherited from the woman, but it would avoid cloning, which, even if it becomes efficient, will always be an heroic procedure.

An overview of the extant gene transfer technologies and their potential for future development would suggest that an acceptable germ line genetic intervention procedure will probably involve cloning. The tissue culture environment allows for addition of genes, removal of genes, and even replacement of genes, but these manipulations require selection and screening for identification of successful gene transfer events. As things look today, the tissue culture environment is best suited to selection and screening. Cloning, although presently a primitive technology, allows for the creation of a new individual with a complement of genes that can be first engineered in tissue culture. Moreover, cloning need not be done from adults. Embryos can be used to generate cell lines that, after manipulation, could be used as donors in a nuclear transfer procedure. Another option is the use of fetal cells in the event that correction of a genetic abnormality in a fetus is the objective. Such circumstances might arise when a couple conceives a child only to find out at the time of genetic diagnosis that the child has genetic abnormalities. The fetus could then be aborted, but cells from the abortus could be used to establish cell lines that could then be genetically manipulated and used to create a new embryo and fetus that is a "clone" of the original.

Of course, if cloning became very efficient, it might be used in attempts to produce new individuals in the absence of genetic manipulation. For example, cloning could be used to generate a conceptus from a child that was mortally injured or afflicted with a fatal disease. It could also be used as an approach to infertility or to allow lesbians to parent a child without genetic input from a man. These applications of cloning and their rationales will be compared with alternative approaches to the same medical objectives a bit later. First, though, it would be useful to outline what people can expect when they undertake genetic manipulation of the germ line. Can we expect to design genetic "super-people, " with low risk of disease, high longevity, good looks, and high intelligence? Can we produce a basketball team with five Michael Jordans? Can we replace loved ones who are dying or recently deceased by cloning? To answer these questions, and to see how genetic manipulation measures up against alternative approaches to the realization of health-related objectives for our offspring, we must first examine the extent to which control of genetic makeup leads to control of the developmental outcome.

What Genes Can and Cannot Do

What features of human development must be remembered so that we do not overestimate the power of genetic manipulation to determine how a child will actually develop? Of course, environment significantly affects the way an organism develops. To resurrect the old "nature versus nurture" argument here would be fruitless, however. All of us know that environment plays some role in development, and all of us believe that genes also exert an important influence. Exactly which of the two factors is most important in determining complex traits such as mathematical or musical ability is the subject of conjecture, and each of us has his/her own bias as to how environment and genetics are balanced. In general, people nowadays have a very deterministic view, believing that genes are by far the most important force in determining even the most complex traits. The problem, of course, is that controlled experiments that can sort out the influences of genes and environment are virtually impossible to set up and conduct. Rare situations such as identical twins reared apart present some opportunities to evaluate these issues, but unfortunately, our methods of evaluation are imperfect. For example, concordance between identical twins reared apart for disorders such as schizophrenia is far higher than the prevalence of the disorder in the general population. Therefore, we infer that genes play some role in this disease. However, concordance is not perfect, and even for serious problems like schizophrenia there is argument over how the disease should be defined. Thus, when reading publications on concordance for schizophrenia between identical twins reared apart, it is not unusual to encounter phrases such as "schizo-affective disorder" or "schizoid personality" to describe a discordant twin that does not have symptoms fitting the formal definition of schizophrenia. These other conditions, assumed to have pathological mechanisms similar to, or overlapping with, those of schizophrenia, are mentioned as a way of pointing out that even though twins may be discordant for schizophrenia, they are in fact much more alike then would be suggested by the simple term "discordant." Of course, such a discussion is indicative of a bias on the part of the investigators that genes play a very important role in the disease, but it also shows that the formal definition of schizophrenia is fundamentally descriptive and not as well defined as a tissue diagnosis of cancer, for example.

These problems are minor, however, compared with efforts to measure intelligence or "talent." Did Mozart have more or less musical talent than Beethoven? Was Einstein a better mathematician than Kepler? Although a formal definition exists for schizophrenia, no such definition exists for mathematical ability, musical ability, or any of the other complex functions of the human intellect. So of course it is impossible to devise a test that would answer such questions.

Before we assess what genes can do to influence various traits it would be useful to briefly review some terms that will be used freely in this discussion. *Genotype* refers to the sequence of the genetic material within the individual, whereas *phenotype* refers to the actual physical form and function of the individual. *Alleles* are different versions of the same gene, and *penetrance* refers to the degree to which an alteration in genotype produces an altered phenotype. Thus the genotype of a person with sickle cell anemia is one characterized at the globin locus by the presence of two sickle globin alleles, one inherited from the father and the other from the mother. The phenotype of this person will be one in which sickle cell disease is present. The sickle mutation is highly penetrant, meaning that individuals with two sickle globin genes will virtually always show the sickle cell anemia phenotype. Another highly penetrant gene mutation is that for Huntington disease, a disorder characterized by fatal degeneration of specific regions of the brain in mid life. Because the Huntington disease mutation is dominant, an individual who has one mutant and one normal allele at this locus (the "huntingtin" gene) will develop Huntington disease. With these reminders we will now proceed to look at the genotype-phenotype relationship, starting with cases in which this relationship is strongest and progressing to examples in which the relationship is not so strong. As we progress through these examples, we will discuss them in the context of the amenability of a particular phenotype to modification as a result of controlled alteration of genotype. If we are going to engineer the germ line it will be to achieve specific phenotypic results, and therefore it makes sense only to discuss genetic modification in this context.

Treating or Preventing Disease by Gene Transfer

The most straightforward genotype-phenotype relationships involve single genes that cause disease and have a very high penetrance. These problems are most likely to be successfully manipulated by a germ line genetic modification approach. The two examples mentioned above, sickle cell anemia and Huntington disease, fall within this category. In the case of an embryo carrying the Huntington disease mutation, a gene replacement procedure that substituted the normal Huntingtin gene for the mutated gene would result in a cure for the treated embryo and would eliminate any risk that the embryo, after developing to an adult, would transmit the Huntington disease mutation to his/her own children. Remember that a strategy for gene replacement that compares favorably with regard to risk, morbidity, cost, and effectiveness does not yet exist. Whereas these exotic approaches to avoidance of Huntington disease are at their earliest stages of development, we already have straightforward, safe, and inexpensive approaches to the same problem—prenatal diagnosis and abortion, or preimplantation genetic diagnosis. However, if we assume that an acceptable gene replacement procedure is available, then correction of the Huntington disease mutation would be a good example of a highly effective procedure that would achieve the desired result in every case. The same goes for the sickle mutation. If an embryo destined to develop sickle cell disease had one sickle globin allele replaced by its nonmutated counterpart, the child would be a carrier of the mutation and, for all intents and purposes, would be cured. Note that in this case there is still a 50% chance that the offspring would pass a sickle allele to one of his/her children, because one mutant allele still remains in every cell.

Now let us consider a less straightforward case. There is a gene, BrCA1, the normal allele of which suppresses formation of various tumors. When BrCA1 is mutated and inactivated, a woman carrying the mutation has an 80% chance of developing breast cancer sometime during her life. Her risk of ovarian cancer is also increased. The risk of breast cancer is so great under these circumstances that prophylactic mastectomy has been undertaken to protect patients with BrCA1 mutations from developing breast cancer. What can we expect from a gene transfer procedure that would correct the BrCA1 mutation?

Under this scenario, an individual who might otherwise have been born with a high risk of breast cancer would have that risk dramatically reduced, and preventive gene transfer would certainly be preferable to mastectomy. However, if this kind of gene replacement were performed on every embryo carrying a BrCA1 mutation, we would be forced to acknowledge that fully 20% of the embryos would be receiving a treatment that they in fact did not need. That is because they are in the group with BrCA1 mutations who would never have developed the disease. Moreover, we can't tell the persons born after BrCA1 gene replacement that they will not develop breast cancer. Although the risk of developing breast cancer with the mutation is high, only about 5% of breast cancers are hereditary. The vast majority of these cancers appear sporadically. So now we have a gene transfer procedure that is likely, but not certainly, to be of benefit, and that does not "cure" the disease. Under these circumstances, the patient might well elect to have safe, cosmetically insignificant surgical removal of breast tissue. Removal of breast tissues in a prophylactic procedure would reduce the risk of all breast cancers, not just those caused by BrCA1 mutations, and could well be less risky and less costly than gene replacement followed by cloning.

Now we will examine an even more challenging case. Recall in our discussion of the human genome project (p. 57), that it has been possible to associate regions of five separate chromosomes with an increased likelihood of Parkinson disease in families that have several members affected with the disease (see p. 123). We presume that as the genome project progresses further, at least one functional gene will be identified in each of these chromosomal regions, alleles of which either increase or decrease slightly the predisposition to development of Parkinson disease. Remember that at the moment "anonymous" DNA fragments in these regions have been found to be present slightly more frequently in persons with Parkinson disease, and that actual genes have not yet been identified.

These five chromosomal regions probably have influence on the development of Parkinson disease, but the association is so weak that uncertainty still remains that the cosegregation of the anonymous DNA markers with the disease is not just a matter of chance. Thus if a gene were found in each of these regions that affected development of the disease, and if alleles were identified that appeared to be present more frequently in affected individuals, we could theoretically perform five gene replacements in a cell line and then use that cell line to produce an individual by nuclear transfer. But when we were done with this heroic, costly, and risky procedure, we would only be able to say that we think the newborn has a reduced risk of developing Parkinson disease. Under these circumstances we have undertaken an invasive, expensive, and hazardous (remember that even when cloning is "perfected," it, like all other medical procedures, will have hazards) procedure that might be of no benefit whatever. We must also realize that no matter how refined gene replacement and cloning technology becomes, it will always have a failure rate. So now a couple might very well choose to simply conceive a child by the usual method. That procedure, sexual intercourse, costs nothing and has a high success rate. Moreover, the process of conceiving the child by this method might even be regarded as having negative morbidity! So here we can say that gene replacement would almost certainly be unjustified. We also must recognize that we will not reach an era in medical science when Parkinson disease is eliminated by genetic manipulation. We will discuss this reality further a bit later.

Treating Disease versus Genetic "Enhancement"

Thus far we have discussed the use of gene transfer for eliminating genetic disease or reducing the risk of disease. As complicated as the issues become when a strict relationship between genes and disease does not exist, the use of gene transfer for enhancement of phenotypes reaches still a higher level of complexity. What do we mean by "enhancement"? Of course, we mean the use of genetic manipulation not for eliminating a disease process but for improving the function of an individual who has a "normal" disease risk. Some have argued that this distinction is more semantic than real. After all, if we could double the intelligence of any child by gene transfer, and that improved intelligence could be passed to that child's future generations, could we not then argue that the unfortunate children who did not enjoy the benefit of genetic manipulation were, from the point of view of their intelligence, genetically deficient? Although this argument can be made, we will not mire ourselves in it here. Suffice it to say that we can distinguish "elective" from "therapeutic" gene transfer. We will base this assertion on the realization that none of the great geniuses in human history, from Leonardo da Vinci to Goethe to Einstein, was produced by genetic engineering. Therefore, there is always a chance that genetic modification for such purposes is entirely unnecessary.

Why do I say that genetic enhancement represents a new level of complexity in the use of gene transfer? Couldn't we produce more tall children who might be great basketball players by simply putting in a gene for growth hormone? I say this because traits that manifest as improved performance, while they may seem simple, are in fact exceedingly complex. To illustrate this point, let's consider the results of efforts to use gene transfer for enhancement of traits in livestock.

One of the first such efforts involved genetic manipulation for improved growth. In the case of livestock, the goal was to produce animals that grew faster and that could be brought to market much sooner. Such an advance could be worth many millions of dollars to the livestock industry. The strategy was exactly the same as for the example of the basketball player given above: improve growth by transferring a gene for growth hormone. This experiment was first attempted in pigs, as gene microinjection was technically simpler in pigs than in cattle.

When transgenic pigs expressing high amounts of human growth hormone were finally made, startling results were obtained. The animals did not grow any faster than their nontransgenic counterparts, and they developed strange-looking snouts. Several factors undoubtedly contributed to the failure of these animals to exhibit increased growth, but among them was the fact that the animals were relatively listless and ate less than normal pigs. Why would a pig with extra growth hormone genes eat less, and why would it not grow more?

The answer lies in the fact that even straightforward-appearing phenotypes such as rapid growth are vastly more complex than disease processes related to one or even several genes. Present-day livestock have been selected for rapid growth for many decades by the oldest method of genetic manipulation, selective breeding. Selective breeding is conceptually a simple technique. When selecting for rapid growth, one simply identifies members of a litter that grow fastest and maximizes their opportunity for reproduction. At the same time, the slow-growing animals are not permitted to reproduce. As the fastest-growing animals are interbred over many generations, with their fastest-growing offspring in turn selected for breeding, one produces a stock that grows far faster and to a far larger size than their ancestors. Of course, this process selects for genes that foster rapid growth. But how many genes are involved in producing this trait?

Of course, the selected livestock produce large amounts of growth hormone. But they also most probably have been selected for hormone receptors that render high sensitivity to growth hormone. There are probably also alleles of genes that cause the animal to channel a high proportion of its ingested calories toward growth rather than toward a high rate of metabolism, and there are genes that foster the development of a robust skeletal structure that will allow the large animals to bear their additional weight. When I say these genes "probably" exist, it is because we don't really know how many genes are involved in the phenotype of rapid growth, nor do we know precisely how they work. It should therefore not be surprising that when a single one of the dozens, perhaps hundreds, of genes that influence growth is selected for germ line modification, the intended result is not obtained.

Another salient example of the complexity of genetic enhancement is provided by efforts to produce the "Arnold Schwarzenegger cow." Experiments with viruses that cause cancer in chickens led to identification of a viral gene called "ski," which seemed to be the key gene in the development of tumors in chickens. When the ski gene was used to produce transgenic mice, the animals expressed the gene in muscle and developed enormous amounts of muscle. This transgenic mouse with huge muscles was affectionately referred to as the "Arnold Schwarzenegger mouse." People involved in the cattle industry were quite impressed by this finding, and set out to produce a transgenic cow with ski. They dreamed of a cow that yielded three times the number of steaks as an ordinary animal. After very many months of work and tens of thousands of dollars spent, a transgenic cow with the ski gene was born. However, this animal did not develop enormous muscles. Although the muscles of the animal did enlarge, the enlargement was not symmetrical, and soon the animal was unable to stand up. Because it didn't move, the transgenic cow's muscles wasted, and moreover, its ability to breed was compromised. The animal was finally destroyed.

The failure to successfully make the Arnold Schwarzenegger cow again illustrates that traits such as large muscle mass are in fact very complex and involve many genes. Manipulation of a single gene could lead to successful genetic enhancement, but in the absence of a comprehensive knowledge of the genes that contribute to a trait, the likelihood of success is low. In fact, since transgenic technology was extended to livestock in the mid-1980s, a successful enhancement procedure has never been developed.

The lesson from these animal experiments is that efforts to engineer enhanced traits into human beings must await a far more thorough knowledge of the human genome and the way in which genes influence development of complex traits. In the beginning of this discussion I said that if major technological obstacles exist that have no apparent solution we would not just take on faith that they would somehow be solved. That is not the case here. With the advance of the human genome project, it is reasonable for us to envision a day when the genes that influence even the most complex traits are identified and characterized. Although an understanding of the genetic control of development thorough enough to allow consideration of human germ line gene manipulation may be several decades away, it could be argued that such an understanding will ultimately be attained. We may have a way to go before we know why growth hormone gene expression in a pig would make it eat less, but we'll probably get there.

If we assume that a thorough understanding of the way in which 40,000 human genes act directly on the embryo, or indirectly through interactions with each other, to orchestrate human development, what then will be our chance of engineering the kind of people that we really want? To consider this challenge, we will first examine complex physical traits and then turn to the even more complex traits that involve intellectual function.

Fortunately for us, Mother Nature has already provided some clues as to the degree of control of phenotype that we may ultimately achieve by manipulating genotype. Identical twins are two genetically identical individuals that develop from the same fertilized egg within the same uterus. If we could efficiently manipulate the alleles of all 40,000 human genes in an effort to produce an individual with certain desired physical traits and we produced two such individuals, they would be genetically quite analogous to identical twins. They would not be as alike as natural twins, because they would have developed from two separate fertilized eggs or embryos reconstructed after nuclear transfer and they would likely have developed in two separate uterine environments. However, we would at least have determined exactly which alleles at every genetic locus were in the two offspring. When considered from this perspective, identical twins are like the ultimate genetic engineering experiment: two individuals with identical alleles at every genetic locus! Accordingly, an examination of identical twins for complex physical traits can be made to see how well two copies of the same genotype produce two individuals with the same phenotype.

For this analysis, let us first consider the complex physical trait, baseball skill. A skilled baseball player must be physically strong and have good reflexes, etc. The baseball player Jose Canseco, recently retired, was endowed with a very high level of baseball skill. He played in the major leagues from 1985 to 2001. Over the course of his career he hit 462 home runs, 22nd on the all time list, and drove in 1407 runs. In 1988 he was named most valuable player in the American League, the highest individual honor a player can receive. He was the first player ever to hit 40 home runs and steal 40 bases in a single season. Many believe he will be elected to the baseball hall of fame. Interestingly, Jose Canseco has an identical twin brother, Ozzie, who was also interested in pursuing a career in professional baseball. Ozzie Canseco's career was much less successful, however. He played only three years in the major leagues, hit no home runs, and drove in only four runs in his entire career! His career batting average was only .200, much lower than Jose's .266, and far too low for an everyday player who is not a pitcher. What does this dramatic discrepancy tell us about the importance of genotype in producing complex physical phenotypes?

First of all it should be recognized that Jose and Ozzie Canseco are not as different in their baseball skills as these statistics imply. If a person is good enough at baseball to play even one game in the major leagues he is very good indeed, and so Ozzie Canseco should properly be regarded as a great athlete. However, a important lesson from these twins is that even if we could manipulate all 40,000 human genes—perform gene replacements at every locus to obtain what we thought was an optimal pattern of alleles for producing a person with a desired complex physical trait, we would achieve sporadic success at best. At the beginning of this book I alluded to the notion of producing a basketball team with five Michael Jordans by use of cloning technology. The lesson of the Canseco story is that we might produce five people with the same genes as Michael Jordan, but we are not likely to get what most of us assume we would get—five legendary basketball players. An additional issue, alluded to above only in passing, was that both Canseco brothers were interested in baseball. The implication of this statement is that they both might not have chosen baseball careers. To entertain the notion that one of the Canseco brothers might have chosen a line of work other than baseball is to acknowledge that genes have much less control over such processes as decision making, a subject we will visit in more detail momentarily. Suffice it to say for now that if we produced five clones of Michael Jordan and allowed them the same choices and freedoms as other children, we might very well get a string quintet!

It would be unfair not to point out that identical twins are often more similar with regard to complex physical traits than are the Canseco brothers. Tom and Tim Gullikson, identical twin tennis players, both had professional careers, and their skills were more comparable. The stark difference between these twins is that Tim Gullikson developed a brain tumor and died in 1996, whereas his identical twin brother is alive and well. Obviously, these twins are quite dramatically discordant with regard to their health status. However, by way of illustrating the prevailing bias that genes control everything, I should mention that when I pointed out the stark difference between the Gullikson brothers to a colleague, she replied: "Well then, if one got a brain tumor and the other did not, they couldn't have really been identical twins" (!!). Another case of twins with very similar physical phenotypes is that of the Van Arsdale brothers, Tom and Dick. These twins both attended Indiana University, and over the course of their three-year college basketball careers differed in point output by only 12 points. As professionals the Van Arsdales both played 12 years and both scored over 10,000 career points. So sometimes duplication of genotypes produces very similar phenotypes.

Before we move on to the still more challenging situation of complex intellectual functions, a couple of facts should be reemphasized. We will never reach a stage in genetic engineering technology where we will be able to manipulate all 40,000 genes in a single procedure. Moreover, when and if we produce multiple cloned individuals from the same cell line, we will not achieve the degree of similarity that exists for developing identical twins. Clones develop from different eggs in different wombs. The women who carry clones will not have the same physiology or hormone status, and they will not have the same eating habits, etc. Therefore, even a highly advanced genetic manipulation procedure will have far less control of phenotype than Mother Nature achieves with identical twins. We are therefore far more likely to see great discordances between clones than between twins, and we will never reach a time when the physician can say to the would-be parents "I will design for you a baseball player who will reach the top of his profession." Those who dream of such a day are truly dreaming.

Engineering of Intellectual Capacity

We will now consider the most daunting challenge for the genetic engineer: control of intellectual development and function. Again, even this most complex of phenotypes appears almost certainly to have some genetic component. After all, humans differ from chimpanzees in only a very small percentage of their total genetic endowment, yet humans are capable of reasoning in ways completely beyond that of chimpanzees. Animal rights activists often assert that advanced nonhuman primates like chimps have intellectual capacities of 3-year-old children, but such comparisons actually have little meaning. There are tasks of abstract thought and expression that a human child can successfully master that cannot be evaluated in any nonhuman species. Moreover, the much-admired ability of a chimp to design a tool—for example, a piece of grass inserted into a termite nest to retrieve termites for food—bears little resemblance to design of tools like supersonic aircraft, which requires cooperative interaction of thousands of people and communication of complex, abstract ideas. There are also no chimps that can paint the ceiling of the Sistine chapel with the kind of skill or intellectual depth exhibited by Michelangelo. Somehow we all know that "The Creation of Bonzo" would never be imbued with the profundity of thought or creativity of "The Creation of Adam." So a small difference in genes can in fact make an enormous difference in development of intellectual capacity. However, the fact that a few genes can have such a dramatic effect on intellectual development does not perforce indicate that the intellect can be controlled through genetic manipulation. The explanation for this apparent paradox can be found by taking a closer look at the human brain.

The brain is a conglomeration of neurons that are interconnected. This seemingly banal characterization should not be mistaken for a lack of respect for the brain's complexity: Neurons communicate with each other, and it is the totality of these communications that manifests as intellectual function (in addition to other functions of the central nervous system). The number of connections, or synapses, between neurons of the human brain is about 10^{14}, or one hundred thousand billion. Consider for a moment a single type of neuron, the Purkinje cell, which resides in a region called the cerebellum. The function of the cerebellum as a whole is to allow for motor coordination: When we reach for our fork and knife to eat that duck a l'orange we don't miss and grab the tablecloth by accident. The cerebellum has about 50 million Purkinje cells, all working to help provide motor coordination. A single Purkinje cell has about 100,000 synapses. If we remind ourselves that the total number of human genes in the genome is about 40,000, we realize that a single Purkinje cell has 2.5 times the number of synapses as there would be genes available to direct synapse formation even if every human gene were devoted to that task. But to produce all Purkinje cell synapses we are actually short of genes by a factor of 2.5 times 50 million, because there are 50 million Purkinje cells. And when we consider the total number of synapses in the human brain, we have about 2.5 billion times the number of synapses as there exist genes available to encode them!

What these numbers mean is that genes cannot control exactly how a brain develops; they can only provide a blueprint for formation of the synaptic network. And, it might not take very many genes to encode a "grand plan" for brain development, which would then proceed independently of strict genetic control. Clearly there is relatively little overlap between the grand plans for formation of the chimpanzee and human brains, but what about the brains of two humans? Obviously, genetic differences between humans of even dramatically different intellectual performance will be very hard to find. After all, it's not that easy to find differences between the genomes of chimps and humans. What must be appreciated is that once the grand plan is laid out, execution of the plan must occur largely by a process of self-assembly. Genes may control the positions, proliferative potentials, and neurochemical characteristics of neurons, but the connections must form as the result of signaling processes that are one or several steps removed from genetic control. Therefore, although there are almost certainly differences in the genetic blueprints of human brains, these differences are subtle and their effects subject to extragenomic influences.

It is my own belief that within the human species' genetic repertoire for brain development, there are some blueprints that predispose more than others to the development of high intellectual capacity. But the likelihood that a "good blueprint" will lead to a "good brain" is unknown. At this point it is relevant to point out the dramatic influence that environment is known to play on brain development.

The brain is one of the latest organs to complete development. By the time a baby is born the heart pumps and the liver and kidneys function, and the lungs assume their function just a few moments after delivery. But the brain is entirely different. Regions of the brain that are needed to ensure survival function very well. For example, the brain can respond to elevations of blood pressure by slowing the heart or releasing hormones that allow salt to be excreted, which in turn lowers blood pressure. However, the activities of the brain that we collectively refer to as "the intellect" are almost completely nonfunctional in the newborn.[CE1] Higher cognitive thought, which involves analysis of visual and auditory stimuli, is absent. The ability to analyze abstract problems, such as whether a short, fat container can hold more or less liquid than a tall, thin one, is also absent. Memory is very limited as well. Of course we all know these things from personal experience. Most people cannot recall anything significant in their lives before the age of two or even older.

The structure of the brain at birth is indicative of this absence of cognitive function. Many parts of the synaptic tree are not completely functional because many of the nerve cables are unable to carry impulses to the synapses. Efficient transmission of signals along the nerve cable is dependent on an "insulation" system that consists of the deposition of the protein called myelin around the nerve cables, or axons. In the absence of myelination nerve impulses cannot be transmitted effectively. Over the first 10 years of a child's life the brain essentially completes its development. I've always had fun making the point in lectures that if 90% of the brain's synapses are functional at birth and the other

10% become functional over the succeeding 10 years, then synapses become functional within those 10 years at an average rate of 100 million per hour!

The fact that the central nervous system is so late to complete its development creates an opportunity for the external environment to exert a significant impact on the developmental process. Experiments with animals indicate that this impact is not only profound but important to proper completion of the wiring scheme. Examples of experiments that illustrate this point involve placement of eye patches on newborn cats. If a patch is placed over one eye of a newborn cat and kept in place for several months, the development of the cat's brain is affected profoundly: The entire pattern of neuronal wiring that connects the optic nerve to regions of the brain that interpret visual input is altered. To put it simply, this procedure, known as monocular occlusion, actually affects the *structure* of the brain. It should be emphasized that monocular occlusion causes no physical damage to the animal, it simply prevents light from reaching the brain through one eye.

These results do not just show that events of brain development that occur after birth are profoundly affected by environment, they indicate that the brain *depends* for its proper development on signals from the external environment. Moreover, there is no reason to believe that this dependence is limited to the visual system. Given the findings from monocular occlusion experiments, it is reasonable to speculate that other functions of the brain, including higher cognitive function and abstract thought, develop at least in part in response to environmental cues. A good way of thinking about the importance of environment in fostering development of higher cognitive function is to consider the case of Ludwig van Beethoven. Of course, we all know that Beethoven was one of the greatest composers in the history of Western culture, and most of us are aware that Beethoven was deaf. Beethoven actually lost his hearing gradually over his youth and became completely deaf in middle age. When Beethoven wrote his famed Ninth Symphony he was unable to hear, and when he first conducted it he did not know the audience was applauding at the conclusion of the piece until he turned around and looked into the theater. That a deaf person could write Beethoven's Ninth Symphony is hard for most of us mere mortals to comprehend. But now let us consider a scenario in which Beethoven lost his hearing when he was one week old. Had this occurred, the Ninth Symphony would undoubtedly never have been written and the great creative musical genius of Beethoven would never have developed. Although this point may seem obvious, it is useful to keep it in mind as we ponder the factors that lead to the development of talents of any kind in any person.

Before leaving this point, it might be interesting to speculate as to what impact monocular occlusion, in the form of unilateral blindness as a newborn, might have had on the paintings of Leonardo da Vinci. Many of Leonardo's paintings prominently employ techniques of perspective to add a sense of depth. The background in the portrait of "Mona Lisa" is an excellent example. Instead of the plain backdrop seen in many portraits, Mona Lisa has behind her a mysterious landscape with detail shown far into the distance. Had Leonardo been able to see from only one eye, he would not have experienced depth perception and, who knows, might have painted more ordinary portraits or become an engineer.

Although examples such as these are extreme, we cannot escape the reality that even small environmental cues most probably affect the development of higher cognitive function. This being the case, we must accept the fact that even a deep understanding of the human genome will probably not provide us with the information we need to produce an individual with a predetermined set of talents. We might be able to select a good blueprint, but the final construction process will depend on influences about which we presently know very little, and about which we are likely to know little for quite some time to come.

Randomness and Development

Although the influence of environment on cognitive development is likely to be extremely complex, it is at least theoretically possible to figure out some of the external forces that shape postnatal development of the brain. However, there are other factors we cannot predict. These are events that occur at random and could have a profound impact on development; some of the dynamics of development might also be inherently random.

What kind of random events are we talking about? Well, of course, there is background cosmic radiation, and it is a matter of chance whether such radiation will interact with the genome to produce significant changes. There is also an error rate in the complex mechanisms of cell division and differentiation. As pointed out in Chapter 3 (see p. 32), DNA copying is imperfect, and errors are made each time the genome is replicated in preparation for mitosis or meiosis. There are also errors of chromosome segregation as cells divide. Some spectacular examples of this latter kind of error might be worth mentioning to underscore the significance of the "randomness factor" in development:

There are examples in the medical literature of identical twins of opposite sex! This occurs when an early embryo that is genotypically male (XY) sustains an error in chromosome segregation such that the Y chromosome is lost from some cells during during early embryo cleavage. This produces a population of cells in the early embryo that contain a single X chromosome, so-called "XO" cells. XO individuals develop as females with Turner syndrome, a constellation of abnormalities that includes short stature, webbing of the neck, and infertility. Nonetheless, these individuals are unequivocally female. After the error in chromosome segregation, twinning occurs. One twin receives only XO cells, while the other receives the unaffected XY cells. The result is identical twins, one an XY male and the other an XO female. There have also been cases of identical twins in which one has Down syndrome (three chromosomes 21 instead of the normal two) and the other is entirely normal. An example of this accident is shown in Figure 36. Although these kinds of errors are rare, we may infer that many more of them occur than are detected. Errors that occur late in development, whether they involve mistakes in chromosome segregation, DNA copying, or simple mutation as a result of background radiation, may affect a subset of cells in an adult. These errors might never be recognized as a de-

Figure 36 Identical twins discordant for Down syndrome. The twin on the right has Down syndrome, whereas her sister on the left does not. This is because of a random error in chromosome segregation during early development followed by twinning (see text).

velopmental disorder, but they may have an impact on development. These random events could of course complicate efforts to engineer specific traits into an individual simply by manipulating the genome.

Although such random or stochastic events intuitively would seem to be only a minor factor in human development, there is another form of randomness that could significantly complicate efforts to control development through genetic manipulation. To illustrate this form of randomness, we consider for a moment a pendulum. If we hold a pendulum a specific distance from its resting point and release it, we will be able to predict its speed at virtually any point in its swing and we will likewise be able to predict the degree to which the arc it travels shrinks with each swing as the pendulum approaches its resting point. The more carefully we calculate the conditions that exist before releasing the pendulum, the more accurate our predictions of its subsequent swings will be. We can calculate the density of the air through which the pendulum travels, we can measure its precise height before release, etc. The more carefully we perform these measurements, the more accurately we will be able to predict the pendulum's behavior once it is released. This makes sense.

Now consider a slightly different situation: a pendulum with a steel ball at the end, suspended over a table with three magnets embedded in the surface. These magnets are in a triangular configuration, and when the pendulum is released each of the three magnetic fields will act on it, affecting the direction of the swing. In this situation, the path taken by the pendulum will quickly become totally unpredictable. Moreover, increasing the accuracy of our measurements—carefully measuring the positions and strengths of the magnetic fields, for example—will not help us predict the path that the pendulum will take. In systems like these, it makes no difference how accurately the starting conditions are measured; very quickly after the system is set into motion its behavior becomes totally unpredictable. In fact, the behavior of systems such as these is so unpredictable that if we choose starting conditions at random, we won't be significantly worse at predicting the system's behavior than we would be if we took measurements with excruciating accuracy. What I am describing here is what we affectionately refer to as a system subject to chaos dynamics.

What processes of embryonic development, fetal development, organogenesis, cell differentiation, and organ function behave according to chaos dynamics? At the present time this question can't be answered because we don't understand the behavior of these systems thoroughly enough to even describe them formally. However, a number of publications have indicated that at least some physiological systems in adults function according to chaos dynamics.

For example, an analysis of heart rhythms after a heart attack has found that these rhythms are inherently unpredictable and indicative of chaos. The investigators who reported these findings further suggested that the inherent unpredictability of cardiac rhythms in such situations might have survival value, because it creates the possibility that any pattern of electrical conduction through the heart might occur. That the option exists to produce any conduction pattern allows the heart the opportunity to produce a pattern that restores a life-sustaining, regular rhythm. This is only one example of many biological systems that investigations have indicated are chaotic.

If chaos characterizes some developmental systems within the embryo, fetus, or child, then these systems are inherently unpredictable. As such, no effort to "preset" the starting conditions—i.e., to manipulate the genome or even the uterine environment, will lead to control of these systems. We may be compelled to face the reality that development cannot ever be controlled to a degree sufficient for development of a clinically reliable genetic manipulation protocol.

What this discussion indicates thus far is that gene transfer procedures might be used in the future to correct genetic problems, particularly those problems caused by single genes with high penetrance. When this technology becomes available, however, it will still have to be compared to simpler, more reliable, and safer methods of accomplishing the same goal. For example, prevention of genetic diseases such as sickle cell anemia, Tay–Sachs disease, or Huntington disease can easily be done by simply screening the conceptus with amniocentesis and performing an abortion if the fetus is affected. For those cannot tolerate abortion, PGD is available. Although this latter approach is not as good as prenatal diagnosis and abortion, it is still far better than any genetic modification procedure that entails gene replacement in cell lines established from embryos followed by transfer of nuclei in a cloning procedure. Not only will the latter approach involve several invasive complex steps, but in the case of recessive disease it will only be necessary in 25% of the embryos obtained. So it does not seem likely that gene transfer will ever compare well with alternative approaches to prevention of even these most straightforward genetic problems. Genetic manipulation might be performed to reduce risk of disease, as in the case of the Parkinson disease example given above, but here the likelihood that the patient will achieve the desired result is much lower, and no intervention at all becomes an attractive alternative. In the case of complex physical traits, we know from identical twins that control of alleles at all 40,000 loci will still not reliably produce the desired result. Where higher intellectual function is concerned, we have no idea which genetic loci are important for providing a good blueprint, nor is it clear that we will ever know. Moreover, as we have seen, a good blueprint provides no certainty of success in designing a genius, because environmental input in the late-developing central nervous system is so important that it can actually affect the structure of that system. Before we move on to a direct examination of the ethics of these procedures, we should discuss two more issues. These are the "law of diminishing returns" and the possible uses of cloning that do not directly involve manipulation of individual genes.

We will first deal with the law of diminishing returns. We now must realize that if cloning and gene transfer technology were more refined, we might be able to design the "ideal genome." That is, we could conceive an embryo, start a cell line from that embryo, and with the knowledge provided by the human genome project, perform a series of gene replacements that would confer upon the cell line a "perfect" set of alleles that would minimize disease risk, maximize the potential for development of outstanding physical and intellectual characteristics, etc. Of course, given our previous discussion of the complexity of development, we know that "designer genes" cannot possibly guarantee a good outcome, but at least we also might know that the genotype of the embryo conceived from such a cell line by nuclear transfer would not have any genetic handicaps. What interferes with this logic is the law of diminishing returns:

Because "only" 40,000 genes provide all of the information needed to direct embryonic and fetal development, as well as maintenance of function and reproduction during adult life, it follows that individual genes can affect many developmental systems. Again, we already know this because of our previous introduction to pleiotropic mutations; mutations that affect multiple developmental systems. However, here we are referring to something slightly different.

Because the total number of genes available to encode the myriad functions of development is so small, we must appreciate that the impact of genes or groups of genes on multiple systems can make design of an "ideal genotype" very difficult or impossible. Perhaps the best way to understand this principle is to revisit the case of sickle cell anemia. Sickle cell anemia is, of course, a serious condition that can cause significant morbidity and shorten life. Thus replacement of all sickle globin alleles with normal alleles would seem like a good idea. However, the sickle globin gene is common because carriers of the allele have significantly increased resistance to malaria. So, although the presence of the sickle globin allele is a liability under some circumstances, it is an advantage under others. Today there are methods of preventing and treating malaria that are more advantageous than introduction of the sickle globin allele into the germ line. However, that an allele of a gene can be good for some things but bad for others should alert us to the realities of genome design under circumstances in which the total number of genes is so few. We may change alleles of five genes on five chromosomes to minimize the genetic risk of familial Parkinson disease, but it is not at all unlikely that such a procedure would increase risk of a different disease, be associated with rapid aging in some organ system, or be linked to other disadvantageous physical or intellectual phenotypes. Although these problems with the Parkinson disease-related alleles are only hypothetical, it must be appreciated that genome design would ultimately be a game of give and take. Optimization of some traits would likely make other traits less desirable. A great deal more must be known about the human genome before the significance of this problem is fully appreciated. However, I believe it likely that if genetic manipulation does reach a high state of sophistication, a "perfect genome" will not be defined and gene transfer efforts will be focused on specific rather than broad goals.

Another issue to address before moving to a direct discussion of the ethical and legal aspects of genetic engineering is the use of cloning for reasons other than direct genetic manipulation. As stated above, cloning alone should be regarded as a form of genetic manipulation, because it results in birth of a child whose genotype is known beforehand. However, our previous discussions of cloning have involved genetic manipulation of cells before use of their nuclei for insertion into eggs. Before moving to a discussion of other uses of cloning, it is important to remind ourselves that parents wishing to manipulate the genome of their child via a cloning procedure would be required to generate embryos that would subsequently be used to create cell lines, which in turn would be genetically modified and then used as donors in nuclear transfer. The reason it was important to generate embryos is that embryos have a genetic contribution from both parents, which was presumed to be an important priority in these cloning efforts.

Several medical applications of human cloning have been proposed that would not involve gene transfer. The most common proposal is that cloning might be used for treatment of infertility. If either partner were infertile, cell nuclei from either partner could be transferred into eggs retrieved from the female partner to produce a child who was genetically identical to the partner from whom the nucleus was obtained. The presumed advantage of this approach over gamete donation (sperm donation in the case of male infertility, egg donation for female infertility) is that the couple would not endure the introduction of genes from some unknown person in the process of gamete donation. Although a successful cloning procedure of this type would treat infertility and would avoid gamete donation, it is itself imperfect. Were the female partner to be cloned, the male partner would bear no genetic relationship to the child. Moreover, the baby girl would be the genetic twin of the mother, which could have undesirable affects on the relationship between the father and daughter. From a genetics standpoint, cloning of the mother would be no better for the father than sperm donation. Moreover, sperm donation is very inexpensive, highly reliable, and associated with minimal morbidity. It is far superior in these respects to cloning, which is invasive, associated with morbidity for the mother, and very costly and will certainly always have a failure rate that exceeds the failure rate of sperm donation. The argument that cloning avoids introduction of "unwanted genes" is also weak. With the human genome project as advanced as it certainly will be by the time cloning becomes refined enough for clinical use, it will be possible to choose a sperm donor for a complement of alleles that could reduce risk of disease that might be inordinately high in the mother or that could confer a genetic predisposition to desirable traits. By the time cloning is ready for clinical use, gamete donation will not be nearly as much a game of "genetic roulette" as it is today. These same arguments hold for cloning of the male partner. When the female is infertile, however, gamete donation involves IVF and is therefore somewhat more costly and invasive than sperm donation. It should be realized, however, that egg donation is associated with a very high pregnancy rate and is a very effective procedure. In my view, cloning will never compare favorably to gamete donation for treatment of infertility.

Another proposed use of cloning is to produce a child that is genetically identical to another child that might be mortally ill or injured. A good example here would be the use of nuclei from a child who is in an irreversible coma after an auto accident to produce another child who looks exactly like the first. Here cloning also compares unfavorably to the conception of a new child by simple intercourse. Any new child produced by cloning would be a completely different person from the first child despite the physical resemblance. To what purpose would parents chose to replace a dying child with a clone of that child? One cannot escape the reality that this costly, invasive approach would only be used because the parents would in some way equate the cloned child with the original. This situation would not likely be beneficial to the clone, who would naturally be expected to actually be like the original child, but who would be a new, unique individual with his/her own unique needs. For the parents to simply conceive another child is inexpensive, highly effective, and associated with no morbidity other than that of pregnancy. Therefore, the use of cloning to replace a child is not logical and is likely to place undue burdens on the cloned offspring. Cloning here compares poorly to natural conception.

Another proposed use of cloning has been to allow lesbian couples to have a child without the involvement of a man. Of course, if either of two partners in a lesbian relationship is cloned, the other partner will not contribute genetically to the child. To get around this problem, some proponents of cloning have suggested that embryos from both lesbian parents be produced by cloning and that these embryos be aggregated in tissue culture (as has been done in mice, see p. 106) to produce a composite human embryo that contained cells from both parents. The child in this case would be a genetic "patchwork" individual, with cells from both its lesbian parents coexisting in one female child.

At present the consequences of genetic mosaicism are not known. This child would almost certainly have an unusual appearance, especially if the parents were dissimilar in appearance. For example, if one parent were of Scandinavian origin and the other of African origin, patches of dark and light skin distributed over the body of the child would almost certainly be present. In addition to its unusual appearance, the child would have two sets of tissue transplantation genes. In mice, this circumstance allows the genetic mosaic to accept skin grafts from either parental donor and this situation has no obvious adverse consequences for the animal. A human mosaic might actually have an advantage as a potential recipient of blood or other organs for transplant, as it would recognize more tissue transplantation antigens as "self" antigens. However, whether mosaicism in the immune system or other complex systems would in fact be hazardous is unknown.

To suggest that lesbians have children via such an elaborate approach is outlandish and indicative of terrible ignorance on the part of those who propose it. Were one member of a lesbian partnership to accept simple, safe, inexpensive, and effective sperm donation in order to conceive a child, she would not be introducing a man into the family. Fatherhood is much more than anonymous donation of a sperm cell! Moreover, lesbian couples who undertook costly, invasive, relatively ineffective cloning as a method of reproduction would be denied the opportunity to have a male child. Why should we assume that lesbians would only want daughters? Proposals such as these indicate that some of us are so fascinated by the notion of cloning that we feel compelled to find some use for it. However, a little further thought should persuade any rational person that the relatively heroic approach of cloning for purposes such as these makes no good sense and exposes participants to unnecessary risks.

Before turning to a direct examination of the legal and ethical aspects of human germ line gene manipulation, it will be useful to summarize what we might predict for the future of this technology. I believe it reasonable to assume that methods for modifying the germ line that are sufficiently safe and effective eventually will be developed. If we examine the problems and unknowns related to this technology at present, we must presume that an acceptable gene transfer procedure is at least a decade away, probably more. I say this because of the fact that the central nervous system is very late to develop and because the data from animals suggest that cloning, which may well be an integral part of an ultimately acceptable gene transfer protocol, is associated with late developmental anomalies. Given these observations, it will almost certainly be necessary to test this procedure on nonhuman primates, for which more complex tests of cognitive function are possible. Studies in primates take time because of the length of fetal development and early postnatal development and because of the relative difficulty in producing large numbers of animals for study. Moreover, these techniques have yet to be applied successfully to a primate. Finally, it will be necessary to follow such test animals for several years to be certain that postnatal development is normal. It is therefore clear that it will be a long wait before cloning and related techniques are responsibly attempted in humans.

It is also safe to predict that over the next several years a much deeper and broader understanding of the human genome will be obtained from human genome project research. Rare genetic diseases will be characterized at the molecular level, and genes and combinations of genes that affect risks for disease moderately or even slightly will be identified. We can also predict with reasonable certainty that methods for detecting genetic abnormalities in embryos will improve and that assisted reproduction technologies will become safer, less invasive, and less expensive.

What we are not likely to see is a sufficient understanding of developmental genetics to allow us to determine, before the fact, the kinds of persons that we might produce after manipulating the genome. There are random forces impacting on development that we cannot control, and the factors that determine whether the Jose Canseco genotype will develop into Jose or Ozzie Canseco are completely obscure. Even less well understood are the factors that affect the development of higher intellectual function. And, as we know for certain from experiments with animals, environmental factors during early postnatal brain development can even affect the structure of the brain.

A reasonable prediction for the future of germ line gene manipulation, then, is that it will likely become an option for preventing genetic disease, especially where single genes cause or prevent disease with high penetrance. When this time is reached, of course, it will be necessary to present treatment candidates with all the options addressing their health concerns. Many less glamorous approaches to prevention of genetic disease may be preferable because of lower cost, lower morbidity, and higher success rate. Refinement of these less glamorous methodologies is certain to occur, and the circumstances in which direct modification of the genome is the best therapeutic option are likely to be rare.

When these circumstances do arise, it will then be necessary to discuss the broader ethical and legal aspects of human genetic engineering. Many have argued that this discussion should take place now, so that the issues will be thoroughly analyzed in time to deal rationally with genetic manipulation capabilities when those capabilities do come along. I cannot agree with this position for several reasons. First, we don't know what genetic engineering strategies will in fact be workable, nor do we know how the workable techniques might limit our engineering strategies. Consider, for example, a situation in which it is possible to replace a gene with another allele but replacement of more than one gene in a cell line is not consistent with generation of a healthy embryo subsequent to nuclear transfer. In this case, the options for genetic manipulation would be far different than if it was possible to replace multiple genes. In particular, enhancement strategies would be severely limited if cell lines were not amenable to replacement of multiple genes. Another problem is that we do not know what our social priorities will be when this technology becomes available. Will the cost of medical care be so high and resources so low that certain "luxury" procedures will be forbidden altogether? Will other important scientific priorities such as global warming or the population explosion divert talent and resources from more esoteric genetic manipulation research? Will advances in our understanding of the human genome make us realize that some of our goals for genetic manipulation are impossible to attain; or contrariwise, will some objectives thought to be unreachable in fact prove to be feasible? Without answers to these questions we really can't know what specifics we should actually debate. As a result, we are left to bandy about the broad philosophical questions concerning the sanctity of the human genome. These issues are of course fun to debate and discuss, but the discussions are not likely to lead to any practical plan of action.

However, the desire to debate these questions seems irrepressible, and therefore we shall proceed in the next chapter to discuss them. Hopefully, we will be able to organize our thoughts sufficiently well to come up with reliable guidelines for dealing with future advances in this area of reproductive science and medicine.

10

WHAT IF?
ETHICAL AND LEGAL
ASPECTS OF GERM LINE
GENETIC MANIPULATION

> The major cause of American Negroes' intellectual and social
> deficits is hereditary and racially genetic in origin and thus not
> remediable to a major degree by improvement in environment.
> —William Shockley (1910–1989), Nobel laureate in physics

So now it is time to consider the ethical and legal aspects of germ line gene transfer on the assumption that this technology will reach a state of sophistication that allows its use in the clinical setting. It is worth repeating before we proceed with this analysis that germ line gene manipulation is currently unethical for the relatively mundane reason that within its repertoire of techniques there is none that passes muster as an acceptable medical procedure. Where direct gene insertion is concerned, there are very few situations in which the technique can be used after it is determined that gene "therapy" is warranted, because the gene transfer procedures are all performed in the single-celled fertilized egg, whereas the earliest time that genetic diagnosis can be performed is at the four- to six-cell stage of development. In addition, animal studies show that the site of foreign gene insertion cannot be controlled, that the level of foreign gene expression cannot be predicted, and that the efficiency of gene insertion is only about 15%. Because preclinical studies demonstrate that gene insertion is inefficient and very unsafe, it would be unethical to perform any such procedure clinically. Gene replacement technology requires that cell lines be used, and thus cloning by transfer of a nucleus from the modified cell line into an enucleated egg would be required. Again, animal studies show that cloning is woefully inefficient, is almost certain to be associated with multiple anomalies of gene regulation, and is very likely to have disastrous consequences such as late pregnancy loss or neonatal death. Given that inexpensive, relatively noninvasive, very safe, and effective approaches already exist for preventing transmission of genetic disease, it would be nothing short of medical malpractice to perform a gene transfer experiment of this kind at the present time. Thus we really have no controversy at the moment: Genetic manipulation is unethical because it does not meet the standards of proper medical practice. But what if these technologies do become refined to the point where they might be considered an option for treatment of genetic disease? What if I'm wrong when I say predictable and controllable strategies for enhancement of complex physical and mental characteristics will almost certainly never be accomplished, and genetic enhancement becomes a reality? What are the ethical and legal issues to be considered then? I've already expressed my view that a discussion of the "deeper" ethical issues surrounding human germ line manipulation should await that technological progress because we do not know what our societal priorities will be at that time and we do not know what our actual capabilities will be in the area of gene transfer. Nevertheless, let's take the plunge.

Where ethics of "morals" is concerned, we must first identify those points of contention that cannot be resolved objectively. Many believe strongly that the human genome is a uniquely precious body of information and, whether created by God or natural selection, it has a special role in defining what is "human" and should be off-limits to manipulation by the hand of Man. This is not an assertion that can be corroborated or refuted by scientific analysis or an objective presentation of facts. Therefore, we can be certain that reasonable people will disagree on this issue. Another relevant point of contention for which no logically airtight resolution exists relates to the status of the embryo and fetus. Clearly there is disagreement among reasonable individuals as to whether an embryo or fetus is a "human being." It would not be useful here to digress from our mission of discussing genetic manipulation to address the issue of abortion and the related questions of the status of the embryo and fetus. However, this question would become relevant to our more immediate problem if a given genetic manipulation methodology involves creating embryos for the sole purpose of producing cell lines that could then be manipulated and used to donate nuclei in a cloning procedure, or, if genetic manipulation is compared as a therapeutic approach to prenatal diagnosis and abortion, or to PGD, which could involve the destruction of embryos or fetuses. Suffice it to say at this point that reasonable people disagree about the status that should be given to embryos or even late-gestation fetuses. Although few people in the United States believe that a late-term fetus should be regarded as anything less than a human being, this is not true in other developed countries, and even the small degree of disagreement in the United States over this question is more disagreement than is seen over the question of whether a newborn baby is a human being. Clearly everyone believes that a baby is a human being, so the fact that even a few people believe that no fetus can be given this status indicates that the question remains within the realm of philosophy and cannot be resolved scientifically. It is important to note in this regard that in the 1973 *Roe v. Wade* decision, in which the US Supreme Court allowed states to restrict late abortion, the Court still allowed the procedure if it was deemed necessary to protect the "life or health" of the mother. The decision did not define "health" as indicating physical health only, and thus a late-term abortion to preserve the emotional health of the mother would be permissible. I make this point here by way of reminding the reader that, contrary to the prevailing belief, reinforced by the complete ban most states place on late-term abortion, the Supreme Court did not attempt to resolve the philosophical issue of fetal status in its decision.

In addition to identifying issues about which disagreements exist that cannot be resolved definitively, it is important to identify questions about which there is no dispute. That is, as we approach the legal and ethical aspects of germ line genetic engineering, we can use as landmarks for our intellectual navigation those moral and ethical precepts about which we all agree. For example, there is no dispute that any genetic intervention that is ever made should conform in its approach and implementation to the canons of medical ethics. We have already discussed the fact that with present-day technologies this principle cannot be adhered to. However, there are other ethical and legal principles that should also be recognized before we attempt to determine whether genetic engineering of humans can be acceptable in our society. First, we should all agree that once a given form of medical therapy is deemed socially acceptable, we should maximize the number of medically ethical treatment options that the patient may choose from to achieve the therapeutic objective. For example, it makes no sense to allow genetic selection but then restrict the methods of performing this selection in such a way that the therapeutic effort becomes more dangerous, less effective, or associated with more burdensome morbidity or cost than is necessary. To impose such restrictions would not be unlike insisting that all coronary artery disease be treated surgically and banning the use of medication to manage the problem. Limiting acceptable therapeutic options can be very dangerous to patients because of individual differences in health status and ability to tolerate one or another form of treatment. For example, in the hypothetical case of coronary artery disease mentioned above, there are many patients who for one reason or another may be very poor surgical risks. To deny these people the option of treatment by medication could be tantamount to passing a death sentence on them. Although we cannot imagine that the treatment options for coronary artery disease would be restricted in this way, it is not at all outlandish to envision situations in which pressure existed to limit certain approaches to germ line genetic manipulation. The reason for the difference, of course, is that germ line interventions may entail the use of procedures such as abortion, and feelings run high when the issue of pregnancy termination emerges.

The debate over "partial birth" abortion provides an example of this mistake. This procedure is used for pregnancies more advanced than 3 months but under 6 months and entails partial removal of the fetus followed by maceration and complete extraction. The most feasible alternative abortion method is surgical opening of the uterus, or laparotomy, and this latter approach is far more invasive and can be dangerous for the woman. It is important to emphasize that the debate over partial birth abortion is not a debate over abortion at this stage of pregnancy. Abortion of fetuses at this stage of development has already been deemed socially acceptable. Thus those in favor of banning partial birth abortion would not ban abortion, they would only force patients to seek another treatment option, laparotomy, that could be more dangerous and associated with greater morbidity. Thus the fundamental principle that the patient should be able to select the best treatment option for a socially accepted therapy is being challenged by this debate. If this debate is won by those who would outlaw partial birth abortion, physicians could be placed in a very difficult position. No physician wants to injure or kill a patient with a procedure such as a laparotomy because an alternative treatment, already shown to be safe and effective, is banned from use. For the doctor to employ less than the best treatment is a violation of the covenant between patient and doctor.

Another principle about which we can all agree is that any procedures or methods to be employed for achieving the therapeutic objective should be developed and practiced in accordance with our best scientific understanding of the problem being addressed. Clearly we do not want "miracle cures" offered to patients when the scientific literature indicates that such cures are impossible. On the other side of the coin, we don't want to restrict patients from selecting a therapeutic approach that the scientific literature shows to be safe and effective. Again, although it is hard to imagine that laws restricting a medical procedure would not be formulated on the basis of reliable scientific information, the principle of scientific validation can be strayed from in this area of reproductive medicine because of abortion. To illustrate this point we can return again to the debate over partial birth abortion. On July 23, 1998, Henry Hyde, a member of the House of Representatives from Illinois, made a speech opposing partial birth abortion. The following is an excerpt from that speech:

> Loneliness. We all know something about loneliness. It is one of life's most mournful experiences. We have all been lonely, and it teaches us how much we humans need each other. What a special loneliness it must be for that little almost-born baby to be surrounded by people who want to kill him.

This utterance would indicate that Rep. Hyde would like to ban partial birth abortion in order to prevent fetuses (male fetuses, at any rate!) from suffering a particularly severe form of loneliness. However, our knowledge of neurological development tells us that it is scientifically impossible for a fetus at this stage of development to experience loneliness—the brain function required to produce such an emotion is not yet developed. Again, because emotions run high when the contentious issue of pregnancy termination is debated, we can stray from the fundamental principle that medical procedures should be developed, implemented, and regulated in accordance with the best scientific information.

Finally, we don't want our regulation of genetic manipulation or any other form of medical therapy to compromise our basic freedoms any more than is absolutely necessary. Where the practice of medicine is concerned, these freedoms include the right of patients to explore and undertake socially accepted therapies in an environment that preserves confidentiality. As alluded to previously, decisions regarding invasive manipulation of one's person require for their optimization a relationship of trust between patient and caregiver that cannot be maintained without confidentiality. As I also stated earlier, laws that tell us what we can and cannot do with our own bodies or organs represent the ultimate form of intrusion on our personal freedom. Therefore, we should strive as much as possible to allow our citizens to maintain sovereignty over their own bodies.

There are some other features of "ethics" as it applies to the practice of medicine that must be kept in mind as we move toward discuss genetic engineering. First, we must face the reality that codes of social conduct are developed not only in accordance with what we view as inviolable principles of "right and wrong" but also with what is practical for our society. Many people flinch when told their ethical and moral codes are developed pragmatically, but it is simply unrealistic to expect any group of people to establish codes of conduct that would do themselves harm. To illustrate this point, consider the story of the airline passengers marooned in the Andes after a plane crash, chronicled in Piers Paul Read's book *Alive*. Most of the crash survivors, Uruguayan boys, were religious Catholics whose beliefs were entirely contrary to the notion that cannibalism is acceptable. However, when trapped in the Andes mountains for 10 weeks with no food and facing certain death from starvation, they managed to reconstruct their moral perspective, convinced themselves that cannibalism was morally acceptable, and proceeded to eat the bodies of victims who died in the crash. This decision was not hypocritical; rather, it was a reflection of the reality that establishment of codes of conduct is a pragmatic process and that it does not lead to development of moral standards to which adherence is highly detrimental.

Another, more immediate example involves the reaction in the United States to the terrorist attacks of September 11, 2001. One of the philosophical cornerstones of American democracy is the principle embodied in the writ of habeas corpus. The right of citizens not to be held by the authorities without charges being lodged or legal counsel being made available is a critically important restraint on potentially repressive behavior by the government and is held as one of the highest moral principles of our democracy. However, in the wake of the September 11 attacks, many Americans were detained by the Department of Justice and held for many weeks without charges being filed and without the opportunity to consult legal counsel. Although some newspapers, citizens' groups, and even international human rights organizations expressed concern over these actions, the general public accepted them fairly well. Presumably this acceptance was based on the belief that the need for security overrode the legal and moral priorities of the writ of habeas corpus.

Yet another clear example of pragmatism affecting principle is legalized gambling. Just a few years ago gambling was considered immoral and was outlawed in almost every state, with Nevada representing the outstanding exception. However, as state budgets tightened and the patronage of numbers games run by organized crime became more candidly recognized, states began running their own numbers. Today, nearly every state has legalized gambling in the form of a lottery.

As we attempt to develop ethical principles to provide a framework for guidelines that define the acceptable limits of germ line genetic engineering, we must keep in mind that these principles must be adopted in the context of what is practical in our present-day society. We don't want to allow procedures that defy our consensus of what is morally and ethically acceptable, but at the same time we do not want to establish restrictions that unnecessarily limit fundamental freedoms or that harm people while doing no good.

Another important fact to appreciate as we move forward is that from the point of view of what is ethical or what should be legal, medicine is a "moving target." When Lou Gehrig died of amyotrophic lateral sclerosis, a disease that causes progressive muscle paralysis and eventual death from respiratory failure or related complications, there were no important ethical concerns over the way in which the disease was to be managed: When people developed this disease they were sustained in the best way possible until they died, which generally occurred no more that 2 years from the time of diagnosis. Today, however, people with amyotrophic lateral sclerosis (now referred to by the sobriquet "Lou Gehrig disease") can be placed in situations that pose ethical challenges. These challenges arise from the fact that, with modern medical technology, an individual who is totally paralyzed can be maintained on life-sustaining equipment for a great many years. A person with end-stage Lou Gehrig disease may be unable to move, eat, or speak, and he or she may find life unbearable. Yet such individuals cannot act to end their own lives, as others with terminal illnesses that don't affect motor function are able to do. Persons in this situation might therefore request assistance in terminating their lives, and we are all aware of the intense debate over the acceptability of assisted suicide. Such situations appear almost daily in newspapers, and the central issues are familiar to most of us. However, there are some wrinkles to this situation that are not often appreciated. For example, there is an active and highly laudable movement in this country to ensure that the disabled are not discriminated against. If a person with Lou Gehrig disease who also has an advanced and terribly painful cancer cannot exercise a decision to terminate life, while others with such terminal conditions but who are not disabled are able to do so, does a ban on assisted suicide discriminate against the disabled? This issue, as well as the more commonly debated questions surrounding the desire of some to seek assisted suicide, arise from advances in medical technology that now allow people to remain alive despite very advanced disease conditions. Thus, although assisted suicide might have seemed ethically unacceptable in Lou Gehrig's time, some states are now enacting laws that permit this procedure.

The "moving target" principle also applies to genetic engineering and related technologies that involve reproductive medicine. Historically, human preimplantation embryos, which contain only a few cells and which float freely in the fluid of the fallopian tube and uterus, have not been accorded any special status. Birth control devices such as the intrauterine device, or IUD, which prevent embryos from implanting into the uterine wall and establishing a pregnancy, have never had their use restricted because they lead to the demise of preimplantation embryos. Similarly, birth control pills that block implantation and "morning-after pills" have not been restricted for the reason that preimplantation embryos are killed by being denied the opportunity to implant. Even the Catholic Church has not historically accorded "human" status to preimplantation embryos. The Vatican has always opposed abortion at any stage, but not because it ascribed the status of "human being" to preimplantation embryos. Rather, early abortion was opposed because it indicated that intercourse had been engaged in for some reason other than procreation, and as such, constituted evidence of "cruel lust." St. Augustine, one of the leading philosophers in the history of Catholicism, once estimated that a human soul was present in the conceptus after 30 days, which is well beyond the time of implantation.

Assisted reproductive technologies have changed this perspective, however. The ability to fertilize eggs in the laboratory and generate preimplantation embryos that develop into babies after return to the uterus has heightened awareness of the developmental potential of preimplantation embryos. Although it has long been recognized that these embryos have the potential to become human beings, the willful discarding of these embryos in cases, for example, where frozen embryos are abandoned by the biological parents, has created more discomfort than existed when the embryos were simply not allowed to implant. Now many people regard these embryos as the equivalent of human beings and the destruction of them as murder.

Continued progress in reproductive and genetic engineering technologies promises to further complicate this situation. If one regards a cell or cells with the potential to develop into a baby as the equivalent of a human being, then cells from adult organs, which have now been shown through cloning technology also to have this developmental potential, might acquire a new and greater status. Think of the complications that would ensue if removal of an appendix were regarded as the willful destruction of millions of human beings! Obviously, this kind of ethical stance is completely impractical, and I have already emphasized that ethical precepts are established, at least to some degree, on the basis of what is practical. However, to dismiss adult cells that, if fused to an enucleated egg, could develop into a baby as not having the status of human beings while regarding the cells of the early embryo as a human being, is not scientifically consistent. I will offer a solution to this problem later. Suffice it to say for now that medical progress is creating a problem that cannot, if we are to strive for philosophical consistency, be easily dismissed.

While we are on this subject, we might further consider how, in the context of present-day medical technology, we define an entity as a human being. Embryos don't look like people at all, but fetuses, as has often been emphasized by opponents of abortion, do have at least a basic human form. If having a human form defines an entity as human, then we certainly cannot discontinue life support for comatose adults, who look much more human than any fetus. In addition, preimplantation embryos look nothing like a human being. We may rationalize our higher regard for fetuses or embryos on the basis that they have the potential to become a person, but again, pursuit of this line of reasoning can lead to serious philosophical difficulties: An entity defined as having the potential to become a human is then, by definition, not a human, and thus this kind of reasoning can actually strengthen the position of those who would favor allowing such entities to be discarded. If we define humans as having certain functional attributes, such as higher neurological function, then comatose individuals with minimal brain activity might fail to be regarded as full-fledged living humans, and discontinuance of life support could be justified. However, this definition would also exclude even late-gestation fetuses. Similarly, if we define a human being by his/her ability to function independent of heroic support as might be used to sustain a comatose individual, then fetuses again would be in trouble: There is no life support system more elaborate or sophisticated than a human placenta!

Abortion laws in most states are predicated upon the notion that after a certain point in gestation a fetus might be able to survive independent of the life support provided by the placenta, if the opportunity to function independently was made available. This notion is clearly attractive, as it forms the basis for most state abortion laws, which prohibit abortions under any circumstances after 6 months of pregnancy. However, the famous case of Karen Ann Quinlan undermines this definition as well. In 1975, at age 21, Karen Ann Quinlan lapsed into a coma after ingesting drugs and alcohol. She remained in what was described as a "persistent vegetative state" on a respirator. After a long legal battle, her family finally succeeded in having her life support removed. But Ms. Quinlan continued to breathe on her own for an additional 10 years, eventually dying in 1985. Now, of course, Ms. Quinlan could not have survived without medical assistance, because she couldn't eat and drink on her own, etc. But babies also cannot survive without this kind of support. Perhaps we may distinguish the persistent vegetative state from fetal status by asserting that people in Ms. Quinlan's condition have no potential to attain full mental function. But many embryos and fetuses also often fail to develop into human beings with high-level neurological function. Moreover, many people suffer developmental problems or traumatic accidents that lead to a state of profound mental retardation. These individuals may well breathe on their own, but they will never develop a high level of cognition. Nonetheless, the profoundly retarded are universally regarded as human beings, and the willful killing of such people is regarded as murder.

Advances in medicine, by creating a new cognizance of the developmental potential of embryos, fetuses, and even adult cell nuclei and by allowing individuals who would otherwise die to survive for many years, are creating, and will continue to create, ethical and moral dilemmas. As we move to regulate genetic engineering technology these realities are best kept in the forefront of our thinking. Of particular importance are issues relating to the status of embryos and fetuses, because, as stated before, genetic engineering can involve the creation of embryos in the laboratory, and treatment alternatives in this area of medicine often include pregnancy termination.

Before turning to specific issues relating to the ethical and legal aspects of genetic manipulation, there are a few more general points about the enactment of laws as a method of regulating medical practice. Again, because reproductive medicine is so important to genetic engineering technology, these issues will be discussed in the context of this specialty.

There are circumstances under which regulation of an activity by legal proscription is highly effective. Laws are particularly useful when they are enacted for the purpose of protecting people from others or even from themselves. Although many people bristle at the notion that they should be protected from themselves by the state, there is ample precedent for enactment of laws that provide such protection and good evidence that these laws are socially beneficial. For example, laws enforcing the use of seat belts in automobiles protect drivers from their own mistakes. Although seat belt laws took some time to be promulgated and accepted, they have contributed enormously to the decline in highway fatalities. Seat belt laws protect no one but the drivers who are required to obey them. Of course, most laws protect us from the actions of other individuals or groups, and we all value these legal restrictions for the function they have to increase our personal security and freedom. Nobody argues that legal punishment should not be forthcoming for an act of murder, for example.

Legal restrictions work less well, however, when they are put in place simply as a method of expressing a moral position. The 18th amendment to the Constitution, which was ratified on January 19, 1919, banned the production or consumption of alcohol. Referred to as the "noble experiment" by Herbert Hoover, prohibition laws were in large measure the result of what was viewed as the morally debauched behavior seen in saloons around the country. Thus legislators felt they were taking the moral high road by legally banning alcohol consumption. Unfortunately, prohibition laws functioned in almost no other capacity than to codify the view of many who believed alcohol consumption to be immoral. The result, of course, was widespread disregard for the law, the emergence of organized crime syndicates that provided illegal alcohol to consumers, and a rapid realization that prohibition laws made things worse rather than better. Consequently, only a few years later, on December 12, 1933, the 21st amendment, which repealed Prohibition, was ratified. Today, laws that restrict the use of alcohol for the purpose of protecting the citizenry, such as application of legal penalties for driving while intoxicated, are widely accepted. As we consider using the judicial apparatus for regulating genetic engineering technologies, it would be wise to recall the history of Prohibition. We don't want people to engage in behaviors that will injure others or perhaps even themselves, but on the other hand, we don't want to make laws against certain behaviors simply as a method of making a statement that we believe such behaviors to be morally wrong.

There are other important ramifications to the enactment of laws that would ban certain medical procedures. It is obviously important that the practice of medicine be legally regulated. We don't want people practicing medicine without a license, we don't want physicians employing deceptive practices that lead to unnecessary procedures or overcharging of patients, and we don't want physicians to practice medicine with egregious incompetence. Laws protecting patients from the improper practice of medicine are critical protections against abuse of the powers physicians have to oversee the physical and emotional health of their patients. Nobody argues against the need for such protections.

However, there are other kinds of legal controls that are not beneficial to patients. These laws do not regulate the quality of caregiving but, instead, the treatment choices available to physicians for addressing a medical problem. An example of this kind of regulation was mentioned above in the discussion of partial birth abortion. If a physician is allowed to perform an abortion at a given stage of pregnancy, it is very unwise to restrict the abortion method for any reason other than effectiveness and safety. Restrictions that limit physician choice for other reasons, for example, to avoid having aborted fetuses experience loneliness, only place the physician in a position where patients could be unnecessarily harmed. An example more immediately relevant to germ line gene manipulation involves discarding of embryos. At the moment, laws are being considered that would ban the use of preimplantation embryos for production of stem cell lines that would subsequently be used for research. Consider a scenario in which a couple undergoes IVF, obtains more embryos than can be transferred back to the woman, and elects to freeze the supernumerary embryos. This couple conceives fraternal twins from the embryo transfer and decides that the embryos remaining in the freezer are no longer wanted. The couple decides not to donate these embryos to other infertile couples and requests that instead, rather than being discarded, the embryos be used for research. The couple states that although they don't want these embryos for conception of any more children, they don't want them "wasted" and that they feel the same way about donating the embryos for research as do parents who wish to donate a dying child's organs for research or transplantation. That is, the couple states that they will suffer significant emotional trauma if the embryos are simply thrown in the waste basket.

In this situation the embryos will surely be discarded, because no one would sanction the removal of the embryos from the custody of the biological owners. However, laws against donating embryos for research would require that the embryos be discarded rather than being used for research. This law harms the couple who are compelled to discard the embryos and ,secondarily, harms the caregiver, who is unable to provide the kind of holistic care that I alluded to above as critically important to the practice of medicine. By restricting the way in which a medical decision is carried out, rather than the competence with which it is carried out, such a law does only harm.

Another important feature of legal regulation of medical practice is that laws are inflexible. Laws "draw a line in the sand" and state that to step over that line is to incur punishment. Again, it is sometimes necessary to establish firm limits on behavior by use of the judicial apparatus. But often problems in medicine are not well addressed by this approach because of the diversity of conditions and circumstances that can affect patients. Although we may have the best of intentions when we limit medical decision making by law, we can inadvertently cause extreme emotional pain to patients while failing to achieve a tangible social benefit. Let us consider how this might happen, again using examples from the field of reproductive medicine, which is so closely linked to germ line genetic engineering technology.

As mentioned previously, many states have used the 24th–26th weeks of pregnancy as the time after which the fetus has potential to live independent of the support provided by the placenta. As a consequence, abortions after this stage are prohibited. Before proceeding to illustrate some of the negative consequences of this immutable restriction, it is worth pointing out again that it draws a rather arbitrary line in the sand. If a woman has an abortion at 23½weeks she and her physician are within the law, but if she has the same procedure just a few days later, a law is broken. Another important point regarding this particular legal restriction is that abortions after 24 weeks of pregnancy are exceedingly rare and almost never performed simply to terminate unwanted pregnancies that result from ineffective birth control. Therefore, in many respects these state laws do not address a social problem—wanton killing of potentially viable fetuses. Rather, they codify a moral position that the killing of potentially viable fetuses is wrong.

Let us now consider a married couple with two children who desire a third child. They conceive the child and do everything that expectant parents should to optimize the pregnancy. However, at 23 weeks of gestation an ultrasound is done, and the fetus is found to have a severe abnormality. This abnormality will not result in fetal death but will certainly lead to death of the newborn just a few weeks after birth. The expectant mother then states that she cannot cope with the experience of watching her newborn baby die. She says that this experience will be so traumatic for her that her ability to be a good mother to her living children will be compromised because of the resulting depression she is sure to suffer. Accordingly, she requests an abortion. However, because of bureaucratic and logistic impediments, she is unable to schedule the procedure until the 25th week of gestation. Under many state laws, this woman would be forced to bear her child and experience its subsequent death. Note that laws disallowing pregnancy termination after the 24th week have not benefited the conceptus, which will surely die, and they have harmed the couple emotionally while at the same time delaying efforts to conceive another healthy baby. If the emotional harm done were to lead to a significant clinical depression or even a suicide by the mother, substantial and irrevocable harm would have been done. For the situation described here there are a few states and several foreign countries that allow late- term abortions. However, not all people have the resources to travel long distances to terminate a pregnancy, and a legal restriction against pregnancy termination should not be justified on the basis that the procedure is allowed in some other jurisdiction. One way patients in such situations can escape legal punishment for undergoing a late abortion without traveling is to lie about conception dates. After all, the estimated day of conception is in fact only an estimate—nobody really knows exactly when the sperm entered the egg. However, a law that encourages otherwise honest people to lie must be of questionable value. We may therefore summarize this situation by concluding that although the law against late-term abortion is well motivated, the inflexibility of the legal mechanism has created a situation in which the very citizens the law is designed to benefit are harmed in exchange for no tangible benefit either to society or to the affected patients.

In situations such as the one described, we may be inclined to think legal prohibition against pregnancy termination too harsh and argue that under circumstances in which the fetus is known to be incapable of surviving, late abortions should be allowed. But now let's make the situation a little bit more complicated:

Consider a young woman who runs away from home to the inner city, where she becomes part of a "fringe" social group. Her associations lead to experimentation with drugs, and soon the girl becomes addicted to cocaine. With no means of income to support her habit, and no job qualifications, she turns to prostitution. As a result of this activity she develops gonorrhea, syphilis, and AIDS. The AIDS infection rages, and she secondarily develops toxoplasmosis, which commonly accompanies AIDS infection. She also develops a vaginal herpes virus infection.

This girl begins to lose weight and become generally sickly, and she stops having menstrual periods. She erroneously assumes her cessation of menstruation is caused by her illness. Only after many weeks go by does she begin to suspect she may be pregnant. Finally, she grits her teeth and visits a hospital emergency room for a pregnancy test. The test is positive, but by this time she is estimated to be 23 weeks pregnant. The girl requests an abortion, saying that the baby will surely be an orphan, because she is dying and the father, being one of her customers, cannot be identified. She doesn't want to give birth to an orphan. Moreover, she realizes that the baby will likely be addicted to drugs and could well suffer blindness from congenital toxoplasmosis. Congenital herpes could affect the baby's neurological development, and of course, the baby could also very well be infected with HIV. We should hardly be surprised that given the many potential problems with the pregnancy and the stress of having a child that will be an orphan, the girl requests pregnancy termination. However, again logistics of scheduling make it impossible for her to have an abortion before 25 weeks of gestation.

This second situation is different from the first in that the baby does have some chance of survival. In this situation we are left to judge whether the emotional stress suffered by the woman faced with the prospect of leaving an orphaned child and the further pain she experiences at the thought of bearing a child with so many health problems are sufficient reasons to allow an abortion. Of course, in states where abortions after 24 weeks of gestation are illegal this question cannot even be discussed. If abortion is illegal it cannot be done.

I do not wish to voice a personal opinion on this scenario, but many people to whom I have presented it strongly assert that this woman should be allowed an abortion. I am not attempting here to resolve the question over whether an abortion should be considered morally acceptable under such circumstances. Rather, I am pointing out that legal prohibition of the procedure creates significant problems for this young woman and for society, which must then do all it can to properly nurture the child (a responsibility that we all know too well will be neglected). If abortion were made a legal option, with the basis of legalization resting on the 1973 Supreme Court decision that late abortion can be permitted to protect the "life or health" of the mother, would we be worse or better off than if such abortions were absolutely prohibited by law? This is the kind of troubling question that legal prohibitions in the field of medicine can raise. My own feeling is that if legalization of abortion at this late stage of pregnancy led to widespread use of late abortion simply as a method of birth control, we might be worse off. However, if such abortions were performed only rarely to address difficult situations such as the ones described here, we might well be better off if we allowed them. If abortions like these are frowned upon but not legally proscribed, we could have a regulatory mechanism better suited to the diverse and individualized problems of human health. Under these circumstances frivolous abortions would be difficult to obtain because most physicians would refuse to do them, whereas abortions could easily be obtained under extreme conditions such as those presented by these two hypothetical cases. The prevailing opinion in American society today appears to be that however painful and problematic laws against late abortion may be, they are valuable because they establish a morally uncompromised position.

As we examine the prospect of legally limiting genetic engineering, it would be useful to keep in mind that we will not likely be able to anticipate every possible medical situation. Therefore, we should be careful to legally restrict procedures only when we are absolutely certain that no circumstances exist that would make the genetic modification procedure acceptable. I will offer some of examples of these situations directly.

There are two other features of the legal mechanism that should be touched on before we proceed to address specifically the legal and ethical aspects of germ line genetic engineering. First, the enactment of a law incurs upon society the responsibility of law enforcement. Laws that are put on the books without an accompanying enforcement apparatus are of course a sham, and failure to enforce a law shirks responsibility. Enforcement of laws that regulate reproductive medicine can be cumbersome. For example, many couples use intracytoplasmic sperm injection, or ICSI, to treat problems of male infertility. ICSI is a highly effective procedure and is perfectly legal. However, the medications, laboratory reagents, equipment, and procedures associated with the performance of ICSI are very similar to those that would be used for gene injection or cloning. Therefore, legal monitoring of infertility clinics might be necessary to enforce laws against gene transfer, and this oversight might include unannounced inspections of such facilities, examination of patient records, or other related monitoring methods. Such activity could be quite unpleasant and intrusive for patients undergoing accepted procedures such as ICSI.

Another important point is that laws cover a finite jurisdiction. Because they cannot regulate or control activity outside their jurisdiction, such legal restrictions would have little effectiveness in the United States if they were not enacted at the federal level. Federal laws would force those seeking genetic manipulation to travel to a foreign country that did not prohibit the procedure, and this burden would be quite significant. The likely result would be limitation of the procedure to the well-to-do, who have the resources to both travel and finance the genetic manipulation procedure.

This discussion is not intended to provide a foundation for arguing against laws that would ban one or even all forms of germ line gene insertion. Rather, my point is that the legal mechanism should be used where and when it will be most effective and will be least likely to harm the innocent. A good example of laws against a medical procedure that work effectively are legal bans against abortion in the final few weeks of pregnancy. These laws codify the moral view held by the vast majority in our society that the fetus would almost certainly survive, and in many instances not even require special care, if delivered should be legally protected against abortion. This law works effectively for several reasons. First, it is highly unusual for major fetal developmental anomalies to be recognized for the first time at such a late stage of pregnancy. Therefore, the problem of the woman who discovers such anomalies at 23 weeks of pregnancy, as described above, does not occur. Second, the law is relatively easily enforced, because a woman at 35–40 weeks of pregnancy cannot be mistakenly thought to be in an earlier stage of gestation, when abortion might be legal. Third, and most importantly, the nearly universal consensus that such abortions would be undesirable in our society makes these laws work. It is virtually impossible to find a physician who would perform such an abortion, and indeed, the social consensus against abortion at this stage is so overwhelming that it is also essentially impossible to find a woman who would seek a pregnancy termination. Moreover, this widespread consensus obviates problems of jurisdictional limitations to the law. Virtually every jurisdiction bans the procedure, so the law does not simply ban a procedure for those lacking the resources to travel to a jurisdiction where the procedure is permitted. So again, laws that regulate medical options are effective when they are easily enforced, widely endorsed by society, and unlikely to cause significant harm.

With this background I will now present my views on how best to handle germ line genetic modification from the ethical and legal point of view.

The Status of the Preimplantation Embryo

Because many genetic engineering procedures, as well as the medically acceptable alternatives to these procedures, will require creation of preimplantation embryos, possible freezing of those embryos, and the discarding and/or destruction of embryos, it is important to establish a principled but ethically practical position on the status of preimplantation embryos. As I mentioned above, embryos at this stage of development, where they contain only a few cells and are floating free in the fluid of the fallopian tube, have never been accorded any special status in our society. Women who wished to use birth control devices that killed such embryos by preventing implantation were never restricted from using those devices because they blocked implantation. Moreover, the fate of the preimplantation embryo historically has always resided within the sole province of the woman who carried it. If she placed a high value on it, as evidenced by her taking measures to maximize the opportunity for implantation and continued pregnancy, then it had a high value. However, if she decided to prevent implantation with an IUD, birth control pills, or morning-after pills, the preimplantation embryo had no value.

However, with present-day IVF technology, and with the future application of this technology to genetic manipulation, the situation is somewhat different. Now we may be creating embryos for the sole purpose of establishing stem cell lines, and we may be willfully discarding embryos from the freezer. How might we revise our assessment of the preimplantation embryo in light of these developments without compromising other principles of individual freedom and privacy?

In my view, the most sensible, logical, and consistent approach is to assign to these cleaving embryos the same significance that they have been given all along but to reconsider issues relating to the genetic "father" of these embryos. Cleaving embryos that float freely within the fluid of the fallopian tube or in tissue culture medium in the laboratory are simply groups of cells with no characteristics that resemble any adult human cells. These cells are not "sentient" any more than other cells. A woman carrying embryos at this stage may be properly regarded as not even being pregnant. After all, if her embryos were cleaving in a laboratory in New York while she lay on the beach in Hawaii we would certainly not regard her as pregnant. So, if the embryos happen to be floating in her fallopian tubes the situation is not substantially changed. In this context, it is reasonable to consider the beginning of true pregnancy as the time at which the embryo and mother become physically linked, which is after implantation. If a woman or a couple places a high value on preimplantation embryos, meaning that they wish to use them for establishing a pregnancy, then every effort should be made to optimize the environment of these embryos in tissue culture or to take good care of them in the freezer. However, if the decision is that the embryos have no special value, then it should be quite acceptable to use them for creation of cell lines or to discard them. This approach offers several advantages, while at the same time avoiding potentially serious problems as follows:

With this approach, the status of embryos would not substantially change from what it has always been. In the area of ethics, consistency is a very reassuring thing, because it tends to reinforce the validity of our ethical precepts. In addition, this approach would allow women or couples to create embryos for the purpose of generating cell lines. Such cell lines could be used for genetic manipulation followed by reproductive cloning, but they could also be used as stem cell lines for cell therapy of diseases such as Parkinson disease or Alzheimer disease. To deny people the use of their own embryos for such purposes by changing the status of the embryos could severely impede the pursuit of potentially life-saving medical therapies. Although the deliberate creation of embryos for production of cell lines might make some of us uncomfortable, we should also feel discomfort at the thought of denying patients these embryos as a source of possibly life-saving therapies for disease. In my view, given that preimplantation embryos have never previously been accorded any special status, the moral sacrifice of denying their use for medical purposes is greater than the moral sacrifice made by deliberately ending their opportunity to implant in the uterus. The latter has been done all along with IUDs and related methods of birth control, whereas the former restriction would establish a new authoritarian approach to the pursuit of health care that could severely impinge upon the well being of patients.

In addition to protecting the freedom of people to pursue the most advanced medical care, this approach preserves alternatives to genetic engineering as a method of avoiding genetic disease. Even if the assumption on which we predicate this discussion—that genetic manipulation will become sufficiently effective and safe to be an acceptable medical option—is correct, this option will almost certainly not be the best one for all patients. Accordingly, the use of PGD, which entails biopsy of preimplantation embryos for genetic testing followed by the discarding of embryos destined to develop genetic disease, may well be a better treatment choice for many or most people. Raising the status of the preimplantation embryo could make PGD difficult to justify for two major reasons. First, the deliberate discarding of embryos, even if they are affected with genetic disease, could be more problematic. Second, questions would inevitably arise as to the status of the cell or cells removed for genetic testing. Recall that at this stage of development a single embryo cell has full developmental potential (see pp. 104–105, Chapter 5). Therefore, when we remove a cell and destroy it for DNA testing, we are destroying an entity with no less potential to become a human being than the group of cells from which it was removed. If we view this problem from a religious perspective, we might say that at the moment of its separation from the embryo proper, the individual cell destined for genetic analysis would have its own "soul," in the sense that, if implanted, it would develop into a genetic twin of the embryo from which it was removed. Because identical twins have distinct souls, we would be forced to conclude that the act of removing a cell from the cleaving embryo would cause the creation of a new "soul." One would expect that those with devout Christian beliefs would regard with dread the prospect of Man creating new souls in the laboratory!

For all of these practical as well as philosophical reasons, it is most reasonable to treat preimplantation embryos no differently from the way they have always been treated: If they are wanted, take care of them, and if they are not, do with them as the patients wish. This approach would also permit the payment of fees to women for the purpose of obtaining eggs that could then be fertilized and used to develop cell lines. Allowing such a protocol would greatly speed stem cell research and perhaps reduce the impetus for germ line genetic manipulation by treating disease with advanced stem cell lines instead. And, again, this posture would maintain important alternatives to gene transfer for treatment of genetic disease.

The essentially unrestricted use or discarding of preimplantation embryos or the cell lines derived from them does present some ethical problems that must be dealt with, however. First is the notion of paying women to donate eggs for creation of these embryos. If we are to assign no special status to preimplantation embryo cells over that accorded to other human cells, then the main problem with this procedure is the risk undertaken by the female egg donor. Again, ovum donation involves hormonal induction of ovulation and surgical egg retrieval, both of which can be risky. Were we to pay women to undergo a procedure associated with a significant risk of injury or death, we would be straining the bounds of medical ethics.

Fortunately, we have experience with the process of ovum donation because women already donate their eggs to infertile women who want to conceive a child but who have no eggs of their own. Ovum donation is a highly effective method of treating female infertility due to ovarian failure. In these cases, the egg donors often receive a fee to undergo the risks of ovulation induction and egg retrieval. Although the procedure has risks, it has proven sufficiently safe to warrant its use for infertility treatment, and thus we can be satisfied that it would be acceptable to use it for creation of preimplantation embryos that could then be used to generate cell lines.

Additional ethical issues are raised by the placement of embryos or cell lines that are derived from them into tissue culture dishes or the freezer, however, and these relate to the "ownership" of the embryos or cell lines. Historically, preimplantation embryos belonged solely to the woman who carried them, because decisions regarding their fate inevitably involved her physical well-being. However, the ability to maintain embryos in the laboratory or freezer for prolonged periods, thereafter to establish a pregnancy, can increase the ownership claims of the male genetic parent. Consider, for example, the situation in which a man and woman produce embryos and freeze them for later use in establishing a pregnancy but in the interim period they separate. After several years the woman takes a new partner and requests that some frozen embryos be replaced so that she can bear children and parent them with the new partner. However, the genetic male parent, her former partner, objects, saying that the embryos were conceived with the understanding that he would be the father of the baby. Similarly, the man could become intimately involved with a different woman and ask that some of the frozen embryos be transferred into his new partner. He could assert that the embryos are just as much his property as they are the genetic female parent's. The kinds of problems created by such scenarios would not be a great deal different if the embryos were used for creation of cell lines that might later be used for reproductive cloning by nuclear transfer.

The best way to deal with these novel situations is to make couples undertaking such procedures aware of these possibilities and to encourage, if not insist, that participating couples reach a formal agreement as to how embryos or embryo-derived cell lines are to be handled in the event of separation. Although this process, which can be linked to consent for the procedure, cannot provide ironclad guarantees that disagreements will never arise, it can acquaint couples with these possibilities and better prepare them for unforeseen changes in their personal lives. Such arrangements are already commonly instituted by IVF programs that offer embryo freezing.

The Status of the Fetus

How might advances in reproductive and genetic engineering technologies affect the status of the fetus? As previously discussed, the fate of the fetus resides entirely within the province of the pregnant woman until at least the 24th week of pregnancy, after which abortion is illegal in many states. We have also mentioned circumstances under which the legal ban against later abortion can cause serious problems for patients who encounter unanticipated late developmental anomalies in their fetuses. The frequency with which late abortions are sought is very low, and thus we might say that the number of people hurt by absolute bans against the procedure is so few that society accrues an overall benefit by banning the procedure, that benefit being establishment of a moral position against terminating the life of a fetus with the potential to live on its own if delivered.

It should be appreciated that the use of genetic engineering technologies, in particular reproductive cloning, could change this balance. Animal experiments have clearly shown that late fetal or even early neonatal developmental abnormalities are common in conceptuses produced by cloning. Although this discussion is predicated on the assumption that cloning technology will improve before it is used in humans, it is unrealistic to assume that the improvements will completely eliminate the kinds of complications we observe today with animal cloning. Therefore, the use of reproductive cloning and/or gene transfer in humans may increase the frequency with which developmental disasters are first detected after the 24th week of pregnancy. Should such events occur more frequently, pressure will be created to relax laws against late-term abortion. Therefore, as the time approaches when gene transfer with or without cloning becomes ready for use, the issue of late-term abortion might require reevaluation, and a decision may be required concerning the acceptability of procedures that increase the risk, even if only slightly, of developmental problems that appear late in pregnancy.

Defining the Parameters for Acceptable Genetic Engineering

With these broader issues addressed, we can now approach the problem of genetic manipulation and set parameters for an acceptable genetic engineering procedure.

In considering this issue it would be useful to distinguish procedures that cure genetic disease from those that attempt enhancement of an individual who has no recognizable genetic disorder. Consider, for example, the use of gene transfer for correction of the Huntington disease gene in a family in which the gene is present. If we could perform a gene transfer procedure that replaced the Huntington disease gene with the normal "wild-type" allele at this genetic locus, we could not only cure the embryo on which the manipulation was performed but could prevent transmission of the disease gene to any future generations in that family. This achievement would be profoundly important from the point of view of the affected family, and it would be a significant application of genetic engineering technology. However, the use of gene transfer to correct genetic disease in cases like these is nothing more than a new strategy for treating a health problem for which other effective strategies already exist and are already in use. In the case of individuals affected with Huntington disease, the family can employ prenatal diagnosis and abortion, or PGD, to ensure that no conceptus with the disease is born. These accepted methods also elimi-nate the disease gene from the family, and if done vigilantly enough, remove the gene entirely from the family forever. Therefore, from an ethical standpoint we might find gene transfer to be a very acceptable approach to genetic disease. No new or advantageous traits are conferred upon the recipient of therapy, and no nonhuman DNA sequences are introduced into the human genome. Again, assuming that genetic modification procedures are sufficiently effective and safe to conform to the canons of medical ethics, their use to treat genetic dis-ease is not highly controversial. We therefore can define such applications of the technology as acceptable.

Things become more complicated when efforts are made to introduce new DNA sequences. Suppose, for example, we determine that for a given individ-ual, the globin gene from the chimpanzee would be more effective in treating a genetic disorder than the normal human gene. So, instead of substituting a normal human gene for an abnormal one, we replace the abnormal gene with the corresponding gene from a chimpanzee. This kind of procedure raises two significant ethical issues.

Insertion of a chimpanzee gene into the human genome may be regarded as a degree of genetic tampering that is unacceptable. After all, you might consid-er a human being with a chimpanzee globin gene as "part chimpanzee." This notion is certain to arouse concern. The most logical way of addressing this problem is to appreciate that the difference between the chimpanzee and hu-man globin genes is simply a few changes in the DNA bases. The gene per-forms the same function, although the exact characteristics of the protein that carries out the function, in this case, oxygen transport from the lungs to pe-ripheral tissues, is not precisely the same. In effect, we can think of the chim-panzee globin gene as a human globin gene with a few point mutations. In-deed, is quite possible to engineer the base sequence of the chimpanzee globin gene into the human DNA sequence by use of recombinant DNA technology (this procedure is called "site-directed mutagenesis" and relies for its success, of course, on the specificity of base pairing, as described in Chapter 3). Would we feel better if we improved the function of a human globin gene by changing a few bases in the test tube than if we simply isolated the gene from a chim-panzee? If the sequence changes we introduce into the human gene result in the chimpanzee globin base sequence, then of course the two approaches yield an indistinguishable result. Therefore, if certain biochemical reactions in the body are altered slightly in character with resultant benefit to the individual, and if these changes are introduced via engineered base changes in the DNA, it makes no difference if the DNA base changes are obtained by chemical alteration of human genes or insertion of corresponding genes from another species. From a philosophical point of view, the person with globin DNA sequences identical to a chimpanzee is not "part chimpanzee." Rather, the person carries a globin gene with a few of its bases altered.

Although the introduction of novel DNA sequences into the human germ line is accordingly not perforce a transgression of our ethical precepts, it can be used in such a way as to pose ethical challenges. A globin gene with a base sequence that is not identical to any human gene sequence might offer superior therapeutic features, but what happens when this gene is passed to a child who does not have a genetic disease that affects globin function? Recall that the individual who receives the new gene as an embryo will likely have his/her own children once he/she reaches adulthood and, if the genetic disease treated by gene transfer was recessive, could transmit the novel globin sequence to a child who received a normal human gene from the other parent. Now we have a situation in which a novel sequence that is not fulfilling any therapeutic function is present in an otherwise healthy individual. This scenario could have two possible outcomes that would pose ethical challenges.

The first potential problem is that the novel sequence, when present in a healthy person, could have adverse effects. That is, we could create a new genetic disease by gene transfer. This circumstance would call into serious question the ethical acceptability of the original gene transfer procedure. Of course, this situation could be addressed by performing PGD or prenatal diagnosis and abortion on the embryos or fetuses of the genetically modified would-be parent, thereby preventing transmission of the novel gene sequence to any of his/her children. However, if this step were necessary, it would detract from the overall quality of the gene transfer approach for treating the original genetic disease: The requirement for prenatal screening and possible abortion or embryo discard would be regarded as a negative feature of the overall therapy and would weigh in favor of choosing an option other than gene transfer for treating the globin disorder.

Another and perhaps more troubling consequence of the presence of novel sequences in an otherwise healthy individual would be phenotypic enhancement. What if this novel globin sequence allowed the individual to achieve superior athletic performance through increased endurance, or to have exceptionally reduced risk of a heart attack when cardiac blood vessels became narrowed because of atherosclerosis? From the point of view of the child who inherited the gene from the gene transfer recipient, these advantages would be a benefit.

But if such enhancement were achieved, ethical issues relating to the accessibility of the therapy would arise. If, through gene transfer, we confer exceptional advantages on an individual—advantages that could be inherited by descendants for any number of generations—and if this form of genetic therapy was available only to wealthy nonminorities, we would be biologically reinforcing a social inequity and effectively denying minorities and/or those with fewer financial resources an opportunity to compete fairly for their own chance to succeed. Regardless of the nature of any genetic enhancement procedure, this problem would loom as a critical one.

Such a problem could be addressed by not permitting "transgenes" to be transmitted to children who received genetic material that carried with it the potential for significant phenotypic enhancement. Gene transfer strategies have been proposed that would allow the elimination of the new genetic material from sperm and eggs, while retaining it in all other cells. However, if these more hypothetical techniques do not prove to be practicable or safe in humans, prenatal screening against transmission of the novel sequences might be required. Such a necessity would reduce the desirability of the gene transfer procedure by forcing the gene transfer recipient to engage in prenatal screening and elimination of embryos and fetuses that carried the new gene. In addition, the restriction, by imposing such requirements for reproduction of the gene transfer recipient, would curtail the freedom of that individual in a way that violates our principles of individual choice. Moreover, such an oppressive approach would almost certainly encourage the gene transfer recipient to flee the jurisdiction where such restrictions were codified into law and have his/her children somewhere else. Therefore, this strategy would not be effective.

Another approach would be to review all such gene transfer procedures and permit them, on a case-by-case basis, only if they are determined to offer no potential for inducing new disease or conferring enhanced characteristics. This strategy is predicated upon the assumption that these potential consequences of gene insertion could be determined beforehand on the basis of our presumed deeper understanding of developmental genetics that would exist at the time such gene transfer technologies become sufficiently safe and effective for clinical use. This design is attractive because it makes gene transfer available as a therapeutic tool while at the same time providing a mechanism for avoiding difficulties associated either with creation of new genetic diseases or conferring phenotypic enhancement. This approach would work best if other options for treating genetic disease remained available. Therefore, this strategy reinforces the principle stated earlier, that the full gamut of medically acceptable options for treating disease should be kept available to patients. It is in this context that our discussion of problems related to abortion again becomes particularly relevant. Prenatal diagnosis and abortion, or PGD with selective discard of embryos, will almost certainly be important alternatives for many gene transfer strategies.

A third approach to this problem is to make gene transfer procedures that have the potential for phenotypic enhancement available to all who seek them. This strategy would be predicated upon the assumption that if true enhancement were possible, no law or other restrictive measure would successfully prevent its use. We may attempt to prohibit gene transfer procedures that confer phenotypic enhancement, but market forces will in any case lead to their use by those with sufficient resources to travel to a jurisdiction where genetic enhancement is not prohibited.

In this context, it is important to recognize that in our society, more costly and elaborate therapies for disease have historically been preferentially extended to white males. In 2002 the National Academy of Sciences issued a publication entitled *Unequal Treatment: Confronting Racial and Ethnic Disparities in Health Care.* This book clearly demonstrates that even after socioeconomic differences between minorities and whites that lead to differences in access to health care are corrected, minorities receive poorer health care and suffer increased mortality. In a 2002 study in the journal *Cancer,* investigators found that minority women who underwent surgery for breast cancer were 48% more likely to have follow-up radiation, the standard of treatment, omitted compared with white women. This difference existed even between white and minority women in the same socioeconomic bracket who had similar difficulties with access to the health care system. In another publication in the same journal in 2002, investigators found that patients treated surgically for colon cancer were more than twice as likely not to receive follow-up adjuvant therapy if their addresses had zip codes in the lowest quartile of per capita income—that is, in economically deprived communities with relatively high minority populations. In a study published in 2002 in the *Journal of the American Medical Association,* minorities and whites in a managed care treatment setting were compared for rates of breast cancer screening, eye examinations for patients with diabetes, use of blood pressure-reducing drugs after heart attack, and follow-up care after hospitalization for mental illness. Blacks were found to be significantly less likely than whites to receive any of these services. The authors concluded that in managed care health plans, "blacks received poorer quality of care than whites" (*JAMA* 287: 1288, 2002). In another study in this journal published in

2001, researchers examined survival statistics of blacks and whites for 14 different cancers, and found that "compared with whites, blacks had an overall excess risk of death" (*JAMA* 287: 2106, 2001) and that this increased risk was not related to differences in the biology of the diseases between races. In a study published in the *New England Journal of Medicine* in 2002, blacks and whites of comparable age and economic status were compared for the quality of care they received after hospitalization for a heart attack. In these matched-patient cohorts, blacks were found to be significantly less likely to receive high quality care and to suffer significantly increased mortality. There are also extensive studies that show that in-hospital care for coronary artery disease is less aggressive for minorities and women than for white males and that this less aggressive care is associated with increased mortality. Finally, in a remarkable study published in the *New England Journal of Medicine* in 1999, actors of different ages, races, or sexes participated in scripted interviews during which they described symptoms of coronary artery disease such as chest pain. These taped interviews were then presented to physicians who were attending meetings or symposia, and the doctors were asked what diagnostic procedures they would recommend to determine the cause of the symptoms. Although the symptoms were the same in black and white men of the same age, physicians were far more likely to recommend cardiac catheterization, the key diagnostic test for coronary artery disease, if the history of chest pain was given by a white actor rather than a black actor. These several studies are specifically cited here to make the point that the finding of discrimination in health care delivery on the basis of race and sex is not anecdotal but reflective of a consistent and broad social trend.

These and many other studies clearly document a sobering fact: Where access to sophisticated, aggressive health care is concerned, there is pervasive discrimination against blacks and women. This discrimination occurs for treatment modalities both in and out of the hospital and for a variety of disease states. Moreover, the discrimination is not based on logistic issues such as the ability to access the health care system or to pay for treatment.

This reality is ethically problematic in and of itself, but it becomes especially important in a situation in which germ line procedures that confer an enhanced phenotype loom on the horizon. First, because minorities are more likely to have lower incomes, legal restrictions against genetic enhancement would lead to de facto discrimination against minorities who seek its benefits. This would be true because the well-to-do could afford to travel to locations where legal restrictions did not exist. The option of making such treatment legally available to all would partially address this problem by eliminating the costly burden of travel for those who sought genetic enhancement, but if genetic manipulation were not paid for by insurance, its high cost would still result in preferential exclusion of minority candidates for treatment. Moreover, as the aforementioned literature review demonstrates, this option would not completely correct a disparity between the races with respect to access to such treatment. We must face the fact that if germ line genetic enhancement were offered to our population under the prevailing health care atmosphere, minorities would have less access to this technology because of their race.

This reality is disturbing enough where its importance to the health and survival of minorities afflicted with heart disease, cancer, or diabetes is concerned, but ethnically based disparities in health care delivery could introduce a new and profoundly greater form of discrimination if they extended to germ line enhancement procedures. This is because genetic enhancement via germ line modification could be heritable and could give the families of its recipients a permanent advantage. Thus people who are disadvantaged because of historical factors could be preferentially and permanently placed at a biological disadvantage. Effectively, the preferential use of germ line enhancement in nonminority populations could cement inequalities in our society.

There are two ways to approach this profoundly disturbing prospect. One is to ban basic research that could lead to genetic enhancement. This approach is fraught with many problems. First, the desire of human beings to understand themselves and their environment through scientific study is an essential feature of the human intellect and is impossible to suppress. Therefore, any effort to block free inquiry would be not only oppressive but futile. If the research is banned in one place it will certainly be done in another. In the United States, where free enterprise reigns supreme, the only available method of banning basic research is to withhold federal funding. If this occurred for studies in the area of gene transfer, the result would almost certainly be pursuit of the work in other countries, possibly with an exodus of talented scientists who wished to pursue the research. Another consequence would certainly be conduct of the research by private entities that could accrue enormous financial rewards from development of a successful genetic enhancement procedure. Another psychological factor that would undoubtedly contribute to unrelenting progress in genetic enhancement technology is that which drove the space race and the arms race. Just as in these examples, the notion that "it's better if we get it before they do" may well push this research along. Indeed, it would hardly be surprising if our need for supremacy in the area of gene transfer were justified by the same time-tested phrase, " in the interest of national security. "

Another problem with a ban on basic research is that it is morally troublesome. To ban research because of the view that our society is not sufficiently advanced to use the results responsibly is to shirk the responsibility we all have to promote the moral and ethical advancement of society. Instead of rolling up our sleeves and getting to work to eliminate inequities in health care delivery, we would, by banning basic research in germ line gene transfer, be protecting our right to continue behaving in a manner that we all recognize as morally flawed. Banning basic research is therefore a cop out and an act of cowardice.

The argument that we simply cannot trust ourselves with this new and powerful tool is also hypocritical. We all believe that great social progress has taken place since early civilizations were first established, and indeed, we often wax sanctimonious about our enlightened, democratic system of government and our degree of individual freedom and empowerment. We cannot express pride in our degree of social advancement and at the same time argue that further progress cannot be made that would allow us to properly use genetic enhancement technology.

For these reasons, our best response to the impending arrival of germ line genetic enhancement is not to use it as an excuse to regress morally by restricting intellectual freedom and promoting the persistence of social inequities. Rather, our reaction should be to use these developments as a stimulus for coming to grips with, and solving, important social problems that still plague us. If this adaptive, progressive response is made, genetic enhancement might actually prove to be socially beneficial.

Gene transfer strategies that are specifically designed to enhance the human phenotype present slightly different problems from those that are therapeutic but might adventitiously lead to enhancement. Consider, for example, a multiple gene substitution protocol that strongly favors the development of a child with exceptional intelligence. If such a protocol were used only to enhance characteristics of individuals who had no preexisting abnormalities, ethical issues relating to discrimination against the biologically disadvantaged could arise. How could we justify enhancing intellectual characteristics in normal individuals while neglecting those with mental disabilities?

Before addressing this problem it would be useful to digress momentarily for a discussion of the meaning of the term "abnormal." Some have argued that human phenotypes such as intelligence or even physical stature are a continuum and that what is "abnormal" lies either in the eyes of the beholder or the eyes of the individual who possesses the trait being evaluated. How do we determine whether a person of unusually short stature is "pathologically short" or simply shorter than average? If an individual is five feet seven inches tall but always wanted to be a professional basketball player, could he/she be viewed, at least in his/her own eyes, as pathologically short? This question is important because it can define an intervention as either "therapy" or elective enhancement, and the argument that phenotypes are a continuum has been used to justify enhancement by asserting that what is enhancement in one person's view may be therapy in the eyes of another.

In medicine, there is a formal definition of "abnormal" where traits such as intelligence or height are concerned. Once the mean statistic for height in a population is determined, "normal" is defined as the 95 percent of the population who are closest in stature to this mean and evenly distributed around the mean. Thus the shortest 2½ per cent of the population is abnormally short, and the tallest 2½ per cent is abnormally tall. Of course, measurement of traits such as intelligence is much more difficult, as has already been mentioned. We will assume for the sake of this discussion that measures of intelligence that are in some way relevant and reliably predictive of higher intellectual function will be developed by the time procedures for enhancing intelligence genetically are devised. It is worth a brief reminder here that Michelangelo or Mozart might well have failed one of our currently used intelligence tests, yet their titanic contributions must be considered as fundamentally intellectual in nature. Thus some more work is needed on these assessment methodologies.

Keeping in mind that genetic enhancement will be essentially a medical intervention that will be costly, invasive, and associated with risk, it would be somewhat problematic ethically to use the procedure purely electively while neglecting those who could benefit from it therapeutically. Again, this application of the technology would establish a fundamental disparity in health care delivery that could be perpetuated genetically. For this reason it is my view that the morally most acceptable way to use enhancement procedures is to also apply them therapeutically. In fact, the preferential therapeutic use of such technologies would relieve many of the problems associated with gene transfer protocols that were deliberately developed for the purpose of enhancing phenotypes.

It is for the purpose of this discussion that the quotation from William Shockley is placed at the head of this chapter. Although Shockley died relatively recently, in 1989, few people today would subscribe to his view that blacks are intellectually inferior as a race and that their inferiority is genetic. However, if this were the case for any group of people, the most appropriate use of technology that enhanced intelligence would be preferentially as a therapy for their genetic intellectual disability. Shockley believed that people who lived in ghettos were there as a consequence of a genetic inferiority that resulted in low intelligence, sociopathic behavior, or both. And, as the quotation at the head of this chapter demonstrates, Shockley did not believe these deficiencies could be corrected by changing the environment. If this were true, then urban slums would be the first place to look to find candidates for therapeutic enhancement of intelligence, regardless of the race to which these people belonged. Again, as with therapeutic procedures that fortuitously confer enhancement, these therapies would be equally available to all comers, regardless of their socioeconomic status. From an ethical point of view, the preferential use of enhancement technology on the "genetically infirm" would be most appropriate. My own view is that a definitive test for genetically based intellectual disability will not find a higher frequency of the disability in urban slums or anywhere else. However, the appropriate use of enhancement technology would be to help those with such a disability, wherever they were found, and regardless of their socioeconomic status.

We have discussed germ line gene therapy and enhancement thus far in the context of using gene insertion to modify biochemical pathways that already exist in human cells. That is, we would modify alleles at various extant genetic loci in order to modify phenotypes. However, it may be possible not only to modify existing biochemical functions but to introduce entirely new ones. This possibility raises novel ethical problems. Consider the following example of the use of gene transfer to introduce new biochemical functions:

Some powerful antibiotics, like neomycin, cannot be taken internally because of toxic side effects. In the case of neomycin, one of the most serious of these toxic effects is kidney failure. However, as previously noted in our discussion of gene transfer technology, bacteria have evolved genes whose products confer resistance to neomycin (see p. 125). In our discussion of transgenic technology, we assumed that we can graft regulatory sequences from one gene to the coding sequence of another, thereby to achieve targeted, tissue-specific expression of the transgene (see pp. 131–132). Suppose we grafted regulators from genes expressed in the kidney to the bacterial gene that encodes resistance to neomycin and introduced the new hybrid transgene into the human germ line. Results from transgenic mouse experiments would predict that the new gene would be expressed specifically in the kidney. Using a protocol of this type we might be able to engineer an individual in such a way as to eliminate the renal toxicity of neomycin, which would in turn allow systemic use of the drug for treatment of infection. Similar strategies might be developed to engineer resistance to a variety of natural and synthetic toxins.

On the face of it, this kind of strategy sounds attractive. If we could engineer the human species to survive more powerful chemical agents used to treat diseases such as cancer or to resist harmful effects of hazardous substances in the environment, what harm would be done?

In my view these kinds of strategies raise new and important ethical issues. The most obvious of these new problems relates to safety. If Mother Nature spent 30 million centuries to evolve a human genome that does not possess some biochemical functions, it is possible that the abrupt insertion of these new functions could have harmful secondary effects. So the first ethical issue arising from the prospect of introducing new biochemical pathways involves the proper practice of medicine, in particular as it relates to protection of patients from harm. Of course, this problem can be addressed with research. Another, perhaps more troubling issue, however, involves the notion that we might produce individuals whose genomes do not correspond to that lexicon of coding functions that we call the "human genome." By creating individuals with genetic functions that have no known counterpart in the human species, we would be moving toward the creation of a new species.

How would we know if we created a new species? The formal definition of a distinct species is one of reproductive isolation. When animals of different species are bred, their offspring either do not survive or are infertile. Thus the lion and the tiger can be bred to produce a viable hybrid, the liger, but ligers cannot produce cubs with either lions or tigers. This observation formally defines lions and tigers as distinct species.

The transfer of novel but simple genetic functions into the human genome, such as genes that encode resistance to drugs or toxins, is not likely to lead to reproductive isolation of the offspring. However, if more elaborate biochemical pathways were introduced, and if the introductions were made reproducibly into specific sites in the genome for many unrelated children, it might result in production of a group of individuals that could not bear their own children with anyone other than other individuals with the same genetic modification. This circumstance would formally define these transgenic individuals as a new species. Before rendering my own opinion of this prospect, allow me to reiterate that an objective examination of the extant gene transfer literature would predict that the capacity to perform such a genetic engineering procedure will never be attained. However, as we are "taking the plunge" and speculating, we will consider the ethical ramifications of such an event.

My own view is that gene transfer procedures that create even the remote possibility of speciation should be disallowed, if possible. The creation of a re-productively isolated group of "people" would likely have very disruptive so-cial impact. Would these individuals receive special treatment, or would they be discriminated against and shunned? If they committed acts that we define as crimes, against each other or against their nonengineered human counter-parts, would laws be applied to them in the same way as to humans? And what if one of these individuals were to be murdered and the killer offered the defense that he/she did not commit murder but rather destroyed a mem-ber of another species, as would be the case if he/she had killed a dog or a horse? Although it seems a bit silly to contemplate such scenarios, we are, in this chapter, giving ourselves license to be silly. Therefore, we can consider these potential consequences of speciation achieved by human genetic manip-ulation.

Because of the profound issues raised by these scenarios (however unrealis-tic they may be), it is my opinion that, assuming that gene transfer technolo-gy reaches a level of sophistication that would allow new biochemical func-tions to be introduced into the human genome, we should not allow gene transfer technology to be used for this purpose. When I say these procedures should be disallowed "if possible," I am alluding back to the discussion of the obstacles to legal regulation presented by medical procedures. Laws must be enforceable, cover jurisdictions that do not simply limit procedures to those able to travel, and they must do minimal harm. In the case of this kind of gene transfer, little harm could be caused by disallowing the introduction of new biochemical functions into humans via genetic engineering. With regard to enforcement and jurisdictional issues, one possible avenue of regulation is via federal agencies such as the Food and Drug Administration (FDA). The FDA is charged with regulating, for the purpose of public safety and health, foods, medicines, and devices that might be used in clinical medicine. Although gene transfer does not involve the use of novel compounds or de-vices, the FDA also regulates research, and these regulations extend both to privately and publicly funded research throughout the United States. Thus, if the FDA were accorded jurisdiction over these gene transfer procedures, an effective method of regulating them without specifically enacting an new law would be available. It is noteworthy in this regard that the FDA asserted reg-ulatory authority over reproductive cloning in a letter of March 2001, basing its argument on the presumption that a human cloning procedure would be considered clinical research. Thus the FDA, either with its present authority or with new regulatory powers endowed through legislation, could have broad regulatory authority that would allow such gene transfer to be restricted throughout the nation.

Before leaving the issue of the prospect of altering the course of human evolution through gene transfer, it is important that we address the assertion that even gene transfer procedures that do not introduce new genes, procedures such as cloning, can alter the course of human evolution. Editorials published in some of the world's leading scientific journals have stated that cloning and related gene transfer technologies in fact give us the awesome new power to control human evolution. How true are these statements, and how moved, either positively or negatively, should we be by them?

To get a good feel for the importance of such assertions, it is essential to understand what we mean by the term "evolution." Evolution may be regarded as the change, over time, of the frequency of genetic alleles in a population, with the change taking place as a consequence of selective pressure. That is, environmental changes favor the survival and reproduction of members of the population that have certain patterns of genetic alleles, whether these patterns are arrived at by random assortment of alleles in the parents' developing sperm or eggs or by new mutations. Although the phrase "survival of the fittest" is commonly bandied about in nonscientific discussions of evolution, it is important to recognize that, according to evolutionary theory, it is reproduction, not survival, that is most important. An individual has no biological fitness if he/she does not pass genes to offspring and high fitness if many offspring are produced. This is true because the transmission of alleles to offspring increases the frequency of those alleles in the population. Over very long periods of time, selective pressures in the environment, with or without other environmental factors that can lead to physical isolation of subpopulations within a species, can lead to emergence of new species that may be similar genetically to their immediate ancestral species but are reproductively isolated and therefore defined as distinct. When Darwin examined the finches on the Galapagos Islands, he noted that although they were very similar throughout the islands, the finches on some islands were distinct from the finches on others of the islands. The selective pressures that existed in the Galapagos favored the evolution of finches, but minor differences in the environments on different islands, when combined with the reproductive isolation afforded by the island environment, resulted in different species of finches arising in different parts of the archipelago. Indeed, this observation was a major factor leading to the development of Darwin's theory of evolution.

Although alleles of genes are generally transmitted to offspring via meiosis and fertilization, new alleles can arise by spontaneous mutation. As mentioned previously, mutations can occur through errors in DNA replication but also can be induced by background radiation in the environment or other random events. If a new allele is found in an offspring, the frequency of alleles at the involved genetic locus is altered by one. Is this evolution?

The rare appearance of a novel allele at a given locus is of course part of the process of evolution, but it is not evolution in and of itself. Only if the frequency of the allele increases over time by natural selection can we attribute its frequency in the population to evolution. In typical large populations of species, one or a few copies of an allele are not biologically significant. Again, significance resides in the effect, if any, that the presence of the new allele has on the reproductive potential of the organism that carries it. This description of the process of evolution by natural selection is quite simplistic and is not intended to provide a comprehensive discussion of evolutionary theory. However, the basic principles can be applied to a consideration of whether genetic manipulation constitutes an example of humans controlling their own evolution, as some have asserted.

Where the impact on evolution of human genetic manipulation is concerned, gene transfer procedures must be understood within the context of human reproduction at the level of the entire species. At present, 11,000,000 children are born every month on this planet. With about 3,000,000 deaths per month, this adds roughly 8,000,000 people to the population every month. Even if we reach a point where 1000 children were born every year from genetic manipulation procedures, this would only be about one one hundred-thousandth of the new arrivals each year. This frequency is so low that the birth of genetically manipulated children would be little different biologically than the appearance of a new mutation. In and of itself, the production of such children does not impact on human evolution.

However, if the genetic manipulation conferred a reproductive advantage on the recipient such that the genetically engineered person had more children than average and passed the new genetic information to more offspring than average, the new genes would be conferring a selective advantage. Would this, then, constitute human control of human evolution?

In this situation, it would technically be true that the genetic manipulation affected human evolution, but from a practical point of view the impact would be negligible. Considering the small percentage of the population that would initially be endowed with the new genes, it would take many thousands of years for the selective advantage conferred by the new genes to result in any meaningful alteration of gene frequency within the human population. Moreover, even a quick examination of the pattern of human reproduction worldwide shows that economic deprivation correlates most strongly with fecundity. It is in the Third World, where the most basic necessities such as clean water and adequate food are often lacking, that reproductive rates are highest. As we define the process of evolution by natural selection, it is these suffering people who are the most biologically fit. Clearly social factors impact on the process of human reproduction far more than genetic factors. Most likely, genetically manipulated individuals would be born in the developed world, where rates of reproduction are relatively low. If these individuals conformed in their behavior to their families and peers, they most likely would have relatively few children and would be regarded as relatively unfit from an evolutionary point of view.

There is one sense in which the claim that gene transfer would constitute control of human evolution has some merit. Let us imagine that transfer of a specific gene or group of genes was found to be highly beneficial and was performed repeatedly. The new genetic information placed into the genomes of the offspring could properly be regarded as an allelic variant. If many children were born with the same allelic variant placed into their genomes by manipulation, then the frequency of this variant within the human population, however low, would be the result of a force other than natural selection. So, in this technical sense, repeated use of the same gene transfer procedure would abrogate the typical evolutionary process whereby allelic frequencies are determined by natural selection.

An Organized Approach to Evaluation of the Acceptability of New Gene Transfer Procedures

With all of these factors considered, how can we judge the acceptability of new genetic manipulation technologies? Recalling that our mission in this book was to provide a road map for evaluating new technologies, we are now in a position to provide that map. We understand the molecular biology, developmental biology, and reproductive biology that provide the underpinnings for human genetic engineering, and we are aware of the principles of medical practice that must prevail if genetic engineering techniques are used in humans. Hopefully, we also have a sense of the ethical and legal issues that bear upon this technology, especially with regard to the close linkage between germ line genetic manipulation and reproductive medicine. Table 7 presents a flow sheet for evaluating a new genetic engineering procedure. The first step in this evaluation should be to eliminate, or defer for later discussion, issues that cannot be resolved by use of logic or presentation of scientific facts. These issues, which reside within the realm of philosophy, may be important, but they cannot help us to determine whether a procedure is acceptable at a more fundamental ethical level. Only if these fundamental requirements are satisfied will it be necessary to consider the broader philosophical problems. Some examples of these philosophical questions are given.[CE1]

In addition, we should not proceed to evaluate a new methodology without establishing in our minds the ethical principles that we wish not to compromise by using a new technology. These principles are both practical and philosophical. From a practical point of view, we do not wish to violate the canons of medical ethics, and from a philosophical perspective, we do not want to violate essential principles on which our free society depends. It should be noted that these philosophical principles again cannot be proven to be valuable by presentation of logic or facts. We can include them in our discussion from the outset, however, because they are universally agreed upon as cornerstones of our democratic society.

With these fundamental issues laid out, we can then proceed to determine whether the new procedure is medically ethical. We have already discussed the factors we can use to make this determination. Obviously, the new procedure must be reasonably safe, effective, and associated with an acceptable level of morbidity. Cost is also a factor, although historically, medical procedures are rarely banned solely because of their high cost. Safety and effectiveness must be established through preclinical studies in appropriate animal models and, where possible, via preliminary safety studies in human beings. A determination of the medical acceptability of the new technique is an essential prerequisite for further analysis. It is worth keeping in mind here that if a procedure is shown not to be clinically acceptable, we are relieved, at least for the moment, of the need to determine whether it violates our broader philosophical view of what is appropriate human behavior. As the flow sheet shows, we need proceed with our analysis only if we determine that use of the technique would not be unacceptably abusive to patients.

If use of the procedure would constitute medical malpractice, it is my view that no laws need be enacted to regulate it but that it might be desirable to pass a congressional resolution expressing the view that use of the methodology

Table 7

A new development in genetic
engineering technology is reported.

Defer for later discussion
philosophical issues about which
reasonable people disagree and
which cannot be resolved by facts or
logic.
Examples:
  The human genome is sacred
  Scientific progress is inherently
    good
  The embryo or fetus is a human
    being

Establish ethical principles to be
preserved as the new procedure is
considered.
Examples:
  The right to sovereign control
    over one's body
  The right to private and
    confidential interaction with
    the physician
  Health care should be equally
    available to all

Determine whether the new
procedure is medically ethical.

Ethical: Keep other treatment
options available and rate the
suitability of the new procedure
against other approaches from the
perspective of effectiveness, cost,
morbidity, and safety.

Not ethical: Take no action or
issue a congressional resolution
stating that the procedure, if used
in humans, would be medical
malpractice.

Reconsider the procedure within the
context of society's moral precepts.

Procedure accepted if:
  Fewer than 90% of the public
    oppose it on moral grounds.
  Principles of medical deliver are
    preserved.

Procedure legally banned if:
  More than 90% of the public find
    it morally unacceptable.
  A reasonable and effective
    enforcement of the ban is
    possible.

would be medical malpractice. These options are shown in the center right box on the flow sheet. An ideal candidate procedure for this approach would be reproductive cloning. As demonstrated in the Chapter 9, reproductive cloning would constitute intolerable abuse of patients, would be extremely costly, and, given the preclinical data from animals, would be unjustifiable. So why not make cloning illegal?

First, there are natural controls against clinical use of such a poor procedure. It would be very difficult for practitioners to obtain malpractice insurance that would cover them for mishaps and also difficult for patients to receive any reimbursement for the procedure from insurance companies. For a doctor to perform a procedure that could involve invasive manipulations of as many as 20 patients (remember the surrogate mothers) would be essentially impracticable. However, a strong desire clearly exists to express opposition to cloning in some formal way. So again, why a congressional resolution instead of a legal ban?

Legal bans have several weaknesses as previously outlined. There are problems with enforcement, and if such bans are not federal there are also jurisdictional limitations to enforcement. Moreover, it is my own opinion that legal proscription is less persuasive a statement against cloning than would be made by a congressional resolution. Moreover, legal bans threaten punishment for use of the procedure, but they provide little rationale or explanation for the imposition of penalties. A congressional resolution, however, would require no enforcement, would cover all jurisdictions, would provide a strong negative incentive for use of cloning, and would carry with it a meaningful rationale—patient safety. Such a resolution would therefore appear far less arbitrary than a law and would in my opinion be more influential than a law. Because it contained a reasonable and meaningful rationale, a resolution might even influence policy outside the United States.

If a procedure is determined to conform to our principles of proper medical practice, we can then return to the philosophical issues. At this point we will be faced with the problem that consensus on these issues will certainly not exist. The question then arises: What degree of consensus should exist before we proceed to ban a procedure on moral grounds? As the history of prohibition illustrates, a very high degree of consensus will be required if any legal ban is to be effective. When Prohibition was enacted, the prevailing view was that drinking was immoral or contributed to immoral behavior. However, the degree of consensus over the issues was clearly not sufficient to make a ban on alcohol consumption either effective or beneficial to society.

In this flow diagram I suggest that at least 90% of the public should believe a procedure to be "immoral" before it is legally restricted on moral grounds. As noted previously, laws against very late-term abortion, which are on the books mainly as a statement that such abortions would be immoral, are effective because of the very high degree of moral consensus. In the case of very late-term abortion, essentially 100% of the public, including health care providers, is opposed to the procedure. As a result, legal bans are both accepted and effective. Another important aspect of this abortion example that should not be forgotten, however, is amenability to enforcement. Very late pregnancies are difficult to conceal and thus, enforcement of a ban is practicable.

On the bottom of this flow diagram, the issues surrounding legal proscription are accordingly summarized. Medically acceptable procedures should be allowed on moral grounds if a high degree of consensus on the moral issues does not exist and, especially in the case of genetic engineering, if the principles of equal access to health care are maintained. Legal bans may be effective and appropriate if there is high social consensus and if an effective enforcement mechanism can be put in place. An important message here is that legal proscription of these or any other medical procedures should not be undertaken except under rare circumstances, and they should certainly not be used unless:

1. They can be enforced.
2. They do not ban a method of achieving a socially accepted therapeutic goal.
3. They do not impinge upon the patient-doctor relationship in a manner that would compromise the quality of health care delivery.

How can we summarize the ethical and moral aspects of genetic manipulation technology? First, we must acknowledge that there exists no law of nature or society that makes such procedures inherently immoral. Second, we must understand that these procedures must be medically ethical before they can be used. Third, the major ethical challenge posed by germ line genetic manipulation is not whether such manipulation constitutes "playing God" or interfering with natural law. Rather, the challenge is to our social conscience: Are we going to meet the challenge of new and provocative technologies by addressing inequities in our society that would allow these technologies to have a harmful impact, or are we going to engage in regressive, and ultimately futile, efforts to prevent scientific progress? It is not biological evolution that is challenged by human germ line genetic manipulation, but social evolution. If the present-day social problems that would exacerbate the negative impact of genetic engineering are addressed, we will have little to fear from the scientific advances. If we have the fortitude and determination to ensure that social progress keeps pace with scientific progress, genetic manipulation is likely to offer some new and highly beneficial options for improving human health.

A BRIEF EPILOGUE: UNDERSTANDING OUR BIASES

Women are simply not endowed with the same measure of single-minded ambition and the will to succeed in the fiercely competitive world of Western capitalism.
—Patrick Buchanan, former special assistant to President Nixon, former Director of Communications for President Reagan, and 2000 Reform Party candidate for president

A remarkable aspect of the current discussion swirling around what is presently the most visible form of genetic manipulation—cloning—is the absence of ambivalent opinion. People are either strongly opposed to cloning or stating they will clone themselves "for fun" and opening cloning clinics. There are very few people who say they're not sure about the ultimate social impact of human cloning, and even fewer people who say they don't care. Although opinions on cloning appear polarized and at opposite ends of the spectrum, the individuals and groups that have offered their disparate opinions in fact have much in common. What could "reproductive conservatives," who oppose all forms of intervention into the reproductive process, especially abortion, possibly have in common with those who would march ahead and undertake a procedure that would almost certainly lead, at least at present, to an increased frequency of spontaneous and induced late abortions? A look into our history might perhaps provide the answer.

Consider the story of Anne Boleyn [1501(?)–1536]. Anne Boleyn was the second wife of King Henry VIII, but her ascent to the position of Queen of England was a difficult one. Henry VIII was already married at the time he took a fancy to Anne Boleyn, and the Holy See did not condone divorce. Henry must certainly have cared deeply for Anne because, after several years of struggle and political turmoil, he broke with the Holy See and annulled his first marriage in order to marry her. The king was quite taken with Anne Boleyn for several years before the marriage was finally consummated, but in marrying her, he had another purpose in addition to the satisfaction of his desires: He wanted to bear a son who would inherit the throne.

Soon after their marriage (or perhaps a bit before!) Anne Boleyn became pregnant, and she gave birth to their first child in September of 1533. Unfortunately for both her and the king, the child was a girl. Anne became pregnant again in 1534 but miscarried after 6 months of pregnancy. The miscarried fetus was a boy. There may have been another miscarriage of a far less advanced pregnancy earlier in 1534. Anne Boleyn was not the kindest of individuals, but she was a good wife and mother to her daughter. Unfortunately, she never became the mother of a son. Unable to father a son with Anne Boleyn, Henry began to complain that his initial infatuation with her was the result of witchcraft. Anne fell out of grace and was ultimately executed in 1536. Her last words were nothing but praise for the kindness of her husband, the king.

Now let's move ahead 450 years or so, to the case of Angela Carder. In 1989, Angela Carder, who had a cancer treated previously and in apparent remission, desired a to have child with her husband and became pregnant. However, at 25 weeks of gestation, she was found to have a lung tumor, an apparent recurrence of the cancer that was thought to have been in remission for 3 years. Ms. Carter requested aggressive treatment for her cancer with chemotherapy and radiation even though she was informed that the treatments posed a risk to her fetus. The George Washington University Medical Center, where she was admitted, did not feel it could perform a cesarean section to save the fetus until 28 weeks of gestation. The decision was made to treat Ms. Carder's cancer according to her wishes.

However, after reconsidering the issue, the hospital sought a determination from the court as to whether failure to intervene on behalf of what they deemed was a potentially viable fetus by performing a cesarean section was legally justifiable. A hearing was held at the hospital, during which Ms.

Carder's husband and family, acting on her behalf, opposed the surgery on the grounds that she opposed it and that she was not likely to survive it. Despite their appeals, the court ruled that Ms. Carder was legally obligated to undergo a cesarean section to which she would not consent. Although Ms. Carder was very ill and near death, she was lucid when told by one of her obstetricians of the court ruling. After being told she would not likely survive the surgery, she refused to consent to the procedure. However, after an emergency telephone appeal made by the hospital, a three-judge panel ordered the cesarean section. The fetus died within 2 hours of the surgery, and Angela Carder died 2 days later. Some other, more recent vignettes are also relevant to this discussion.

Jennifer Johnson was a 23-year-old cocaine addict in 1989, when she gave birth in Seminole County, Florida. She had sought treatment for her drug addiction while pregnant but was unable to obtain it. On delivery the baby was not noted to have any serious abnormalities, but Ms. Jones advised her doctor that she had used cocaine while pregnant. Urine from both mother and child subsequently tested positive for cocaine. Ms. Johnson was subsequently charged with the crime of delivering cocaine to a minor. She was sentenced to 1 year of house arrest and 14 years' probation.

On May 15, 1999 Regina Denise McKnight of South Carolina delivered a stillborn baby girl. The baby's blood tested positively for by-products of cocaine. Ms. McKnight was tried and convicted of homicide by child abuse and sentenced to 12 years' imprisonment.

In 1992 Cornelia Whitner of South Carolina was convicted of "unlawful child neglect" after it was determined that she had smoked crack cocaine while pregnant. She was sentenced to 8 years in prison, and in 1997 the Supreme Court of South Carolina upheld the conviction, stating that pregnant women who engage in activities that could harm their fetuses were constitutionally subject to prosecution under the state's child abuse laws.

These few out of many related incidents spanning 450 years point to an important fact that is an essential element of any discussion of procedures that, like genetic manipulation, require the use of advanced reproductive technologies. In my view, these events indicate that, over several centuries, we have been imbued with a deeply held bias: that women, although readily acknowledged to be human beings, also serve an important function as "incubators" of our future children. As such, they do not have the same right of sovereignty over their own bodies as men.

I believe these deeply held biases should be distinguished from an overt discriminatory or bigoted view. When doctors suggest less aggressive treatment for the symptoms of heart disease when those symptoms are described by black actors as opposed to white actors, those doctors are not likely to be overtly racist or members of the Ku Klux Klan. On the contrary, many of them would certainly regard themselves as enlightened on the issue of racial equality. So the problem is not that our society formally endorses or codifies the view that the races are unequal; rather, the disparity in the way doctors would manage black and white heart patients is the result of subtle differences in perception inculcated over decades and even centuries. This is not to say that these differences in perception should not be classified as a form of racial bigotry. Indeed, those victimized by such prejudices are not inclined, nor should they be, to be understanding of the weaknesses of those who would discriminate against them. So, when William Shockley insisted that he was not a racist but, rather, simply confronting biological realities honestly and candidly, we should not excuse his bigotry on the grounds that it was nobly motivated. Bigotry is still bigotry. It is my view that similar social forces contribute to the perception of women as vehicles for bringing babies into the world and the belief that where this mission of reproduction conflicts with individual rights, it should supercede those individual rights in priority and social importance. Regardless of the high moral values that are invoked to justify this form of discrimination, it is still discrimination.

The view of women as incubators appears to have been more freely acknowledged in the time of Anne Boleyn, who made no protest when she was about to be executed. But today we pride ourselves on thinking that all people, not just all men, are endowed with inalienable rights. How can we reconcile our claimed enlightenment with actions taken by the state against Angela Carder, Jennifer Johnson, Denise McKnight, Cornelia Whitson, and countless others? The answer is simple: We don't regard women as "human beings" in the same way as we regard men. In most states, after 24–26 weeks of pregnancy, a woman actually loses complete decision making control of a part of her body—her uterus. Effectively, the state assumes control of that organ after this point of pregnancy and can charge the woman or her doctor with a crime if medical intervention is undertaken that would terminate or perhaps even harm a pregnancy. It is a truly remarkable thing that Angela Carder was not allowed to refuse consent for surgery! As noted in our previous discussion of the principles of medical practice (see p. 158) it is the greatest of intrusion on one's freedom when the state assumes control of an organ system, as it did in the case of Ms. Carder.

This bias that women do not have the same right of sovereignty over their physical bodies as men is not malicious or even conscious. Rather, it is the product of a deep and longstanding perception in Western culture of the woman as incubator. As the prosecutions of women who take drugs while pregnant show, the fetus is regarded with equal or even higher priority than the "incubator." So, Angela Carder was most probably sent to an earlier death on the presumption by a judge that her fetus had a better chance than she did of long-term survival. This is true even though, as discussed earlier, the question of whether a fetus is a human being resides within the realm of philosophy, but there is no question whatever that Angela Carder was a human being.

Were we fully cognizant of the enormous imposition such attitudes and laws make on the freedom of women, the entire atmosphere surrounding reproductive decision making would be different. Those who adamantly oppose the notion that a woman should choose what to do with her own uterus would be expressing their deepest sympathy to women who were not allowed to terminate a pregnancy or who would be unable to refuse consent for surgical procedures such as a cesarean section. Yet, from the anti-choice community, we are much more likely to hear expressions of anger and even contempt for women with unwanted pregnancies. It's not that we've forgotten that women are not regarded as entitled to the same rights over their physical persons as men, we simply haven't realized that any government intrusion into such private affairs goes against our own deepest belief in the importance of individual freedom. We haven't realized it because we haven't recognized that women aren't really incubators who owe their organs to society. They are simply human beings with the same needs and right to seek self-fulfillment as men.

It is for this reason that remarks such as those issued by Patrick Buchanan, quoted at the head of this chapter, are so important. Mr. Buchanan has risen to some of the highest positions in our government, and he was just a stone's throw from being president of the United States. Yet he freely acknowledges his belief that women are in many aspects inferior, and as such, not entitled to the full range of opportunities offered by our society.

Such an open acknowledgment by so important a person is a good thing, because it allows us to understand ourselves better. The common reaction to such statements as Mr. Buchanan's is that they are outrageous and that no reasonable person would really agree with them. But this response is self-delusion. Mr. Buchanan is a powerful, well-known figure who is respected by many and who received many votes for president. So, instead of dismissing his comments as "far out" we are far better served by reacting with serious introspection, in an effort to determine not if, but where and how strong, such thoughts stir within ourselves.

In this context it would be useful to return for a moment to the legal restrictions against late-term abortion. If a woman is two weeks away from delivering a baby, nearly everyone agrees that she should not be allowed an abortion. However, a view of this situation that was genuinely free of bias would acknowledge that by disallowing such abortions we are curtailing the freedom of women to control their own bodies. This curtailment may be justified and morally proper in our view, but it is still a curtailment. Thus, even in this straightforward situation we should recognize that we are sacrificing an element of freedom for the women who are restricted in this manner from reproductive choice. We don't hear expressions of regret and compassion for women in such situations, however rare they may be, because of our deeply held biases.

That we have not achieved this state of self-understanding in this area is reflected very well in the current discussion over germ line genetic manipulation, exemplified by the debate over cloning. Those strongly opposed to cloning have generally very conservative views on reproductive freedom. These people not only oppose cloning but are also most often opposed to a woman having the option to choose whether or not to continue a pregnancy. They are swayed by the notion that a woman has an important role as an incubator, and this bias manifests openly as expressions about the sanctity of life. On the other side of the argument are those who are quite enthusiastic about forging ahead immediately with cloning. Indeed, some of these individuals claim to be actively attempting to be cloning a human being. These individuals express little concern for the pain and distress the biological or surrogate mother might suffer if a pregnancy disaster occurred, as the results of animal research predict it would. They view the woman as an incubator also—an instrument for producing the much-desired "first cloned human." Very much less prominent in this debate are expressions of concern for the women who will be undergoing invasive procedures and carrying high-risk pregnancies that could end in physical and/or psychological devastation.

If we were able to exorcise our deeply ingrained views of women as vessels for carrying fetuses and bearing children, those who oppose reproductive free choice would be hard at work devising medical and legal strategies for minimizing the degree to which the policies they support impinge on the freedom and individual rights of women. And those who believe that the new genetic engineering and associated reproductive technologies augur a new era of highly powerful and beneficial strategies for improving human health would be energetically cautioning against the irresponsible premature use of procedures that could seriously hurt the women involved, and they would be lobbying strongly for research that would first make these technologies safe.

If we can understand how our deeply rooted biases are affecting our outlook on human genetic engineering, we can guide this scientific effort down a path that would lead to humane and possibly highly beneficial advances in medicine. If efforts are made toward achieving this understanding, we will have in any case greatly benefited our society, regardless of the technological limitations of genetic engineering that we ultimately encounter. New and powerful technologies, whether they appear in the area of medicine or some other science, present important social challenges. The challenge posed by advances in reproductive and genetic engineering technologies is to ensure that they are used to increase, rather than lessen, the freedom and well-being of all.

INDEX